MODULAR SERIES ON SOLID STATE DEVICES

VOLUME II
The PN Junction Diode
Second Edition

MODULAR SERIES
ON SOLID STATE DEVICES
Robert F. Pierret and Gerold W. Neudeck, Editors

VOLUME II
The PN Junction Diode
Second Edition

GEROLD W. NEUDECK
Purdue University

ADDISON-WESLEY PUBLISHING COMPANY
READING, MASSACHUSETTS
MENLO PARK, CALIFORNIA • NEW YORK
DON MILLS, ONTARIO • WOKINGHAM, ENGLAND
AMSTERDAM • BONN • SYDNEY • SINGAPORE
TOKYO • MADRID • SAN JUAN

This book is in the
Addison-Wesley Modular Series on Solid State Devices

Library of Congress Cataloging-in-Publication Data

Neudeck, Gerold W.
   The PN junction diode.
   (The Modular series on solid state devices; v. 2)
   Includes index.
   1. Diodes, Semiconductor.   I. Title.   II. Series.
TK7871.86.N48 1989   621.3815'22   88-6277
ISBN 0-201-12296-0

Reprinted with corrections February, 1989

16 17 18 19 20-CRS-97

Foreword

Solid state devices have attained a level of sophistication and economic importance far beyond the highest expectations of their inventors. By continually offering better performing devices at lower cost per unit, the electronics industry has penetrated markets never before addressed. A necessary prerequisite to sustaining this growth initiative is an enhanced understanding of the internal workings of solid state devices by modern electronic circuit and systems designers. This is essential because system, circuit, and IC layout design procedures are being merged into a single function. Considering the present and projected needs, we have established this series of books (the Modular Series) to provide a strong intuitive and analytical foundation for dealing with solid state devices.

Volumes I through IV of the Modular Series are written for junior, senior, or possibly first-year graduate students who have had at least an introductory exposure to electric field theory. Emphasis is placed on developing a fundamental understanding of the internal workings of the most basic solid state device structures. With some deletions, the material in the first four volumes is used in a one-semester, three credit-hour, junior-senior course in electrical engineering at Purdue University. The material in each volume is specifically designed to be presented in 10 to 12 fifty-minute lectures.

The volumes of the series are relatively independent of each other, with certain necessary formulas repeated and referenced between volumes. This flexibility enables one to use the volumes sequentially or in selected parts, either as the text for a complete course or as supplemental material. It is also hoped that students, practicing engineers, and scientists will find the series useful for individual instruction, whether it be for reference, review, or home study.

A number of the standard texts on devices have been written like encyclopedias, packed with information, but with little thought as to how the student learns or reasons. Texts that are encyclopedic in nature are often difficult for students to read and may even present barriers to understanding. By breaking the material into smaller units of information, and by writing *for students*, we have hopefully constructed volumes which are truly readable and comprehensible. We have also sought to strike a healthy balance between the presentation of basic concepts and practical information.

Problems which are included at the end of each chapter constitute an important component of the learning program. Some of the problems extend the theory presented in the text or are designed to reinforce topics of prime importance. Others are numerical problems which provide the reader with an intuitive feel for the typical size of key parameters. When approximations are stated or assumed, the student will then have confidence that cited quantities are indeed orders of magnitude smaller than others. The end-of-chapter problems range in difficulty from very simple to quite challenging. In the second edition we have added a new feature — worked problems or *exercises*. The exercises are collected in Appendix A and are referenced from the appropriate point within the text. The exercises are similar in nature to the end-of-chapter problems. Finally, Appendix B contains volume-review problem sets and answers. These sets contain short answer, test-like, questions which could serve as a review or as a means of self-evaluation.

Reiterating, the emphasis in the first four volumes is on developing a keen understanding of the internal workings of the most basic solid state device structures. However, it is our hope that the volumes will also help (and perhaps motivate) the reader to extend his knowledge — to learn about the many more devices already in use and to even seek information about those presently in research laboratories.

Prof. Gerold W. Neudeck
Prof. Robert F. Pierret
Purdue University
School of Electrical Engineering
W. Lafayette, IN 47907

Contents

3 The Ideal Diode Volt–Ampere Characteristic

4 Deviations from the Ideal Diode

5 P-N Junction Admittance

Introduction

The *p-n* junction diode is the most fundamental of all the semiconductor devices, and for this reason we devote an entire volume to it. So basic is the theory of operation that many engineers have stated that to understand the *p-n* junction qualitatively and quantitatively helps one to understand the majority of all solid state devices. This common thread of application from device to device builds confidence and competence as one progresses into the more exotic device structures. For example, the bipolar transistor is two very closely spaced *p-n* junctions; the solar cell is a special *p-n* junction made to absorb sunlight; the silicon-controlled rectifier (SCR) and triacs are multiple *p-n* junctions; the MOS and junction field effect transistors also contain *p-n* junctions. The list could be continued, but the main point is that a thorough study of the *p-n* junction is a direct path to the other solid state devices, including all of the recent work in VLSI integrated circuit layout and design. Most of the concepts and many of the equations developed for the diode will apply in a very direct and straightforward way to these other devices.

The goal of this volume is to build a firm foundation in *p-n* junction theory from a conceptual and mathematical viewpoint. This will enable the reader not only to appreciate the diode, but — and this is even more important — also to have the necessary background for understanding other present and future solid state devices.

This volume begins by discussing how *p-n* junctions are fabricated in single crystalline semiconductors. The wafer processing steps used in the fabrication of the diode are also basic to the production of other solid state devices such as the bipolar and MOS transistors used in integrated circuits. Chapter 2 details the internal workings of the junction, in particular the depletion region, the region between the *P* and *N* material. The case of thermal equilibrium is contrasted with the cases of forward and reverse bias. Once all the electrons and holes have been accounted for, the charge density, electric field, and potential are developed. The case of forward and reverse bias for the ideal diode and the accompanying volt–ampere (*V–I*) characteristic are examined in Chapter 3. This chapter is extremely important for understanding the concepts of the device physics involved in many devices. Chapter 4 is devoted to deviations from the ideal diode (i.e., the real world of): avalanche breakdown, generation–recombination, and high-level injection. Chapter 5 discusses the small-signal admittance, conductance, depletion capacitance, and diffusion capacitance which limit the frequency response and are critical in linear IC designs. Chapter 6 develops the large-signal switching transient for turning the diode "off" abruptly when forward biased and the turn-on transient when reverse biased. This is important to digital logic IC's. Chapter 7 explains the workings of the Schottky barrier diode that is made of metal directly on silicon. Here the application is to make faster diodes that are mainly majority carrier devices. Also included in this discussion are metal–semiconductor ohmic contacts which are used to make an electrical connection to the bulk silicon.

1 / Introduction to the Diode

The *p-n* junction diode is produced by forming a single crystal semiconductor such that part of the crystal is doped *p*-type and the other part is doped *n*-type.[†] Junctions are classified by how the transition from the *p*-type to *n*-type is developed within the single crystal. The junction is said to be *abrupt* when the transition is extremely narrow. The *graded* junction is a junction where the transition region is "spread out" over a larger distance. In later chapters we will indicate how the nature of the junction's "abruptness" affects the electrical characteristics.

1.1 *p-n* JUNCTION FABRICATION TECHNIQUES

Abrupt *p-n* junctions are formed by alloying a solid impurity (metal) with the semiconductor or by epitaxial growth of silicon directly on a silicon substrate. Graded junctions are produced by gaseous diffusion or by ion implantation of impurities into the substrate semiconductor.

1.1.1 Alloyed Junctions

The most ideally abrupt of the *p-n* junctions results from a fabrication process called *alloying*. In the alloying process a p^+-n junction is formed by starting with an *n*-type semiconductor wafer, placing on the surface of the wafer a doping impurity (often a metal), and heating until the impurity reacts with the semiconductor. Figure 1.1 illustrates this process for *n*-type silicon and the doping impurity of aluminum (Al) or indium (In). Remember that aluminum and indium act as acceptors in silicon. When heated to an alloying temperature of 580 °C the Al softens and "eats up" some of the silicon below the molten Al; that is, it forms a solution of Al and Si atoms. When the temperature is carefully lowered, the silicon begins to regrow on the atomic sites of the host *n*-type silicon wafer. However, the number of Al atoms present in the Al–Si solution is far

---

[†]A more detailed presentation of *p-n* junction fabrication can be found in Volume V of this series, *Introduction to Microelectronic Fabrication*, by R. C. Jaeger.

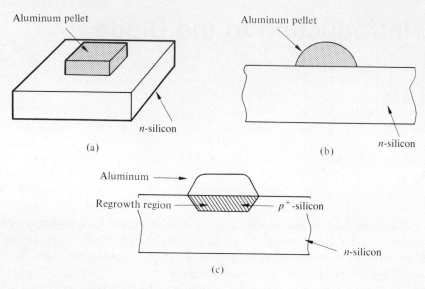

Fig. 1.1   (a) Alloying Al to n-silicon; (b) cross section of (a) after melting; (c) cross section of (a) after melting and cooling.

greater than that of the n-type impurities and the regrowth region is changed to a heavily doped p-type region. In fact, it is degenerate p-type, or p^+. The maximum number of Al atoms possible in the silicon, under normal conditions, is called its *solid solubility limit* ($\simeq 3 \times 10^{18}$/cm^3 Al in Si at 580 °C). The net result of the alloying process is a p^+-n junction with a very sharp (abrupt) transition between the p-type and n-type impurities. As illustrated by Fig. 1.1, the remainder of Al, that which was not incorporated into the silicon, is left at the surface and is a convenient electrical contact to the p^+ region. Also note that the Al has selectively formed the surface geometry (area) of the p^+-n junction.

1.1.2   Epitaxial Growth

Epitaxial growth of a semiconductor layer on top of a single crystal semiconductor substrate is another method by which an abrupt junction can be formed. This technique is very common to bipolar integrated circuits. Epitaxial growth is accomplished by heating the host wafer, say n-type silicon, and passing a gas containing silicon tetrachloride (SiCl$_4$) and hydrogen (H) over the surface at a controlled flow rate. The gases react and deposit silicon atoms on the surface of the host wafer. Because the temperature is usually in excess of 1000 °C, the deposited Si atoms have sufficient energy and mobility to align themselves properly to the host-wafer crystal lattice. This forms a continuation of the lattice up from the original surface. Typical epitaxial-layer growth rates are $\simeq$1 micron per minute.

Impurity atoms, in the form of a gaseous compound, can be added to the carrier gas during epitaxial growth to form either n-type or p-type layers. Typical dopants are diborane (B_2H_6) for p-type layers and phosphine (PH_3) for n-type silicon layers. Starting with an n-type host wafer (substrate) and growing a p-type epitaxial (epi) layer forms a fairly abrupt p-n junction. Other combinations are, of course, possible, such as growing an n-type epitaxial layer on a p-type substrate.

The epitaxial process is used extensively in making bipolar and CMOS integrated circuits (IC's). The p-n diode formed in the "epi" process is kept reverse biased and provides for device and circuit isolation from the substrate (host wafer). Epitaxy has also been used in the formation of SOS structures, where SOS stands for silicon–on–sapphire or silicon–on–spinel. Spinels are various mixtures of MgO (magnesium oxide) and Al_2O_3 (aluminum oxide) and are closely related to sapphire. To make a long story short, doped silicon is epitaxially grown on substrates of sapphire or spinel. The incentive for this procedure is the exceptional insulator quality of the sapphire and spinel substrates in isolating circuits in IC designs requiring high-speed devices, especially large-scale integrated circuits (LSI). Other techniques, such as silicon–over–insulator (SOI), are currently under development.

1.1.3 Thermal Diffusion

Junctions in which the transition from p- to n-type silicon occurs over many atomic spacings are called *graded junctions*. An extremely important method, instrumental in the growth of the semiconductor industry, is that of gaseous (thermal) diffusion of the impurities directly into the host crystal. The impurities are introduced into an inert carrier gas and passed over the surface of the silicon wafer. Due to the high temperature and the large number of impurities at the surface, the impurity atoms migrate (diffuse) into the crystal. As one might expect, the impurity distribution (concentration) is largest near the surface and is progressively smaller further into the crystal. The impurities are typically distributed in a complementary error function or as a Gaussian function (more on this later).

An important property of silicon is the ability of its natural oxide, silicon dioxide (SiO_2), to form in an oxidizing atmosphere. SiO_2 is a glass, and hence it is very impervious to moisture and other contaminants. It also serves as a barrier to the diffusion of the desired impurities and therefore allows precise geometric control of the p-n junction area. Figure 1.2 illustrates how the oxide is used to define the surface geometry of a diffused (graded) junction diode.

So important are the properties of Si–SiO_2 that without them the entire computer industry would not even exist as we know it today. Silicon dioxide has enabled us to build integrated circuits in large volume inexpensively; without it our systems would be limited in reliability and complexity and prohibitive in cost. More will be said on this topic as each device is discussed in later chapters and volumes.

During diffusion, the concentration of impurities at the surface imposes a boundary condition on the diffusion process. If the concentration of impurities at the surface is

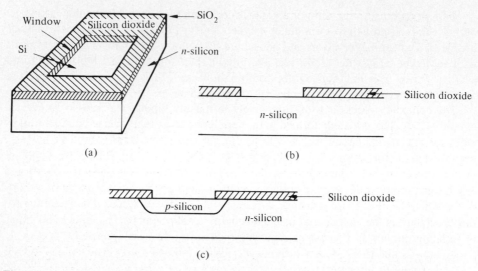

Fig. 1.2 (a) Window in SiO; (b) cross section before diffusion; (c) cross section after *p*-impurity diffusion.

kept constant at N_0 impurities/cm^3, then the impurity distribution is approximated by a complementary error function as pictured in Fig. 1.3(a). There is an unlimited source of impurities at the surface. Starting from the surface the impurity concentration decreases going into the semiconductor. The complementary error function is just another of those unusual functions whose value is often obtained from plots, numerical integration, or tables. Note that the erfc(x) = [1 − erf(x)]. Equation (1.1) lists the mathematical form of the function.

$$N(x,t) = N_0 \, \text{erfc}\left[\frac{x}{2\sqrt{Dt}}\right], \qquad (\#/\text{cm}^3) \qquad (1.1)$$

where

D is the diffusion constant and is a function of temperature and dopant type (cm^2/s) ,

t is the length of time the diffusion takes place (s) ,

N_0 is the surface concentration (#/cm^3) .

The diffusion constant for impurities increases exponentially with temperature. In Fig. 1.3(a), note that as the diffusion time or temperature increases, the impurities move further into the substrate. In addition, the total number of impurities in the sample will also be increased.

It should be obvious that when the diffused impurities outnumber the host (background) impurities, "compensation" has occurred and the host semiconductor is con-

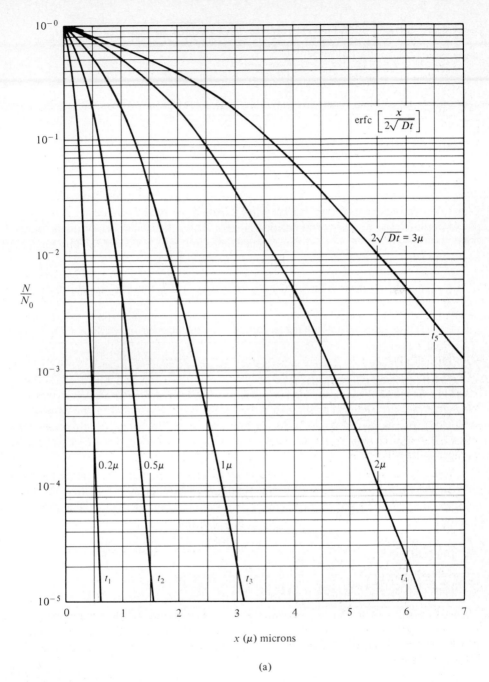

x (μ) microns

(a)

Fig. 1.3 (a) Complementary error function.

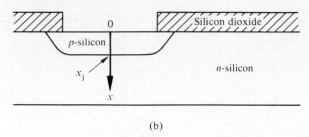

(b)

Fig. 1.3 (b) x-coordinate axis for impurities.

verted to the other type. The p-n junction is formed at a distance below the surface where $N_A = N_D$, as indicated in Fig. 1.3(b). This is called the *metallurgical junction* (x_j).

When the total number of impurities is limited to a fixed number (limited source), then their distribution is approximated by a Gaussian function as described by Eq. (1.2) and illustrated in Fig. 1.4.

$$N(x,t) = \frac{Q}{A\sqrt{\pi Dt}}e^{-x^2/(4Dt)}, \qquad (\#/cm^3) \qquad (1.2)$$

where
Q is the total number of impurities initially deposited near the surface $(\#)$,
A is the area (cm^2),
D is the diffusion constant, a function of temperature and dopant type (cm^2/s),
t is the time of the diffusion (s).

Note that the surface concentration (where $x = 0$) is not constant and decreases as the diffusion time increases. At a distance from the surface the impurity distribution approaches that of an exponential, as evidenced by the nearly straight-line portion near the bottom of Fig. 1.4.

At the junction, the impurities are changing as a function of distance; hence the name *graded junction*. Figure 1.5(a) illustrates the ideal graded junction impurity profile for p- impurities diffused into an n- substrate. This is to be contrasted with the abrupt junction of Fig. 1.5(b). The ideal abrupt junction is also illustrated and from now on will be called the *step junction*. A step junction is a good mathematical approximation to the abrupt (real) junction.

SEE EXERCISE 1.1 – APPENDIX A

1.1.4 Ion Implantation

Ion implantation is a technique for introducing impurities into the single crystal substrate by direct bombardment. Impurity atoms are ionized and then accelerated in a large electric field to energies in the range from 1 to 300 KeV and shot (implanted) directly into

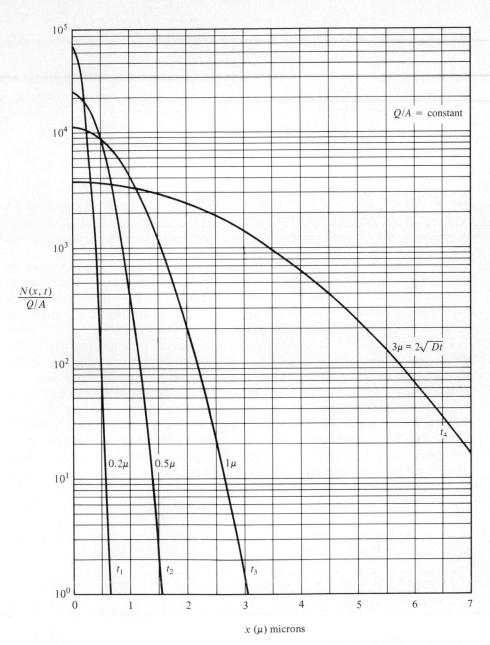

Fig. 1.4 Normalized Gaussian distribution of impurities.

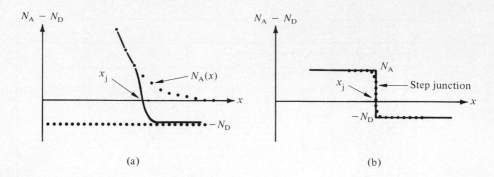

Fig. 1.5 Impurity profiles: (a) diffused *p*-impurities into uniform *n*-substrate; (b) abrupt *p-n* junction · · · · · · ·, step junction ———.

the crystal. The accelerating potential controls the depth to which the impurity ions are implanted, and the current–time product determines the total number of impurity atoms. Needless to say, the crystal is damaged by the colliding ions, but with a thermal anneal (heating up the semiconductor) this damage can be repaired, leaving the impurities activated and the crystal lattice generally intact.

The advantages of ion implantation are precise impurity control, a low-temperature process, and that for some of the III–IV and II–VI semiconductors it is the only reasonable method of doping. Ion implantation also holds great potential for future device development since in principle almost any atom can be implanted into any given substrate. Figure 1.6 illustrates several implants with a range of accelerating voltages.

Very shallow ion implants of impurities are used as a precise "dose" for further thermal diffusion into the crystal. The constant source diffusion is approximated by the Gaussian function, Eq. (1.3). This is a very common procedure in present-day fabrication of small (shallow junction) bipolar and MOS transistors.

$$N(x) = \frac{\phi}{\sqrt{2\pi}(\Delta R_p)} e^{-\frac{[(x-R_p)]^2}{[2(\Delta R_p)]}} \quad (\#/cm^3) \tag{1.3}$$

where

ϕ is the dose ($\#/cm^2$),

ΔR_p is the projected straggle or standard deviation (cm),

R_p is the projected range (cm).

It should be noted from Fig. 1.6 that the straggle (ΔR_p) and projected range (R_p) are a function of the accelerating voltage (ion energy) and the type of ion being implanted. Crystal orientation and substrate material also affect these parameters.

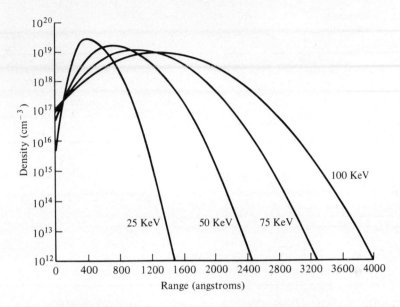

Fig. 1.6 Phosphorus-implant profile in silicon at a constant dose of $10^{14}/cm^2$.

1.2 PHOTOLITHOGRAPHY FOR GEOMETRY CONTROL

The previous section indicated methods by which the surface geometry of the p-n junction could be controlled. The most universal of the methods is that of using a mask which acts as a barrier to the impurity atoms, while openings or *windows* are made where impurity penetration is desired. For example, the thermal diffusion of impurities into silicon through a window in the SiO_2 is illustrated by Fig. 1.2. One might ask the question, "How can one make a window in the SiO_2 2.54 μm on a side?" One ten-thousandth of an inch is 2.54×10^{-4} cm = 2.54 μm. The answer is by using a process called *photolithography*.

Photolithography makes use of an old process known as "photoengraving," which was developed to make very small machine parts. With the scale greatly reduced and aimed at engraving windows for IC manufacturing, the process has become quite sophisticated. Figure 1.7 illustrates the steps used in producing a window in SiO_2 and serves as only one example of the many possible approaches in present use.

A silicon wafer is placed in an oxidation furnace at about 1000 to 1200 °C in a steam and/or oxygen atmosphere to grow a few thousand angstroms ($\approx$5000 Å) of silicon dioxide (SiO_2) on its surface. The gases furnish the oxygen component and the silicon wafer furnishes the Si atoms. This results in the SiO_2 using up some of the silicon wafer surface. When finished, the once metallic gray surface of the silicon will have a range of colors depending on the thickness of the silicon dioxide. This is the reason pictures of IC's are multicolored, with regions of blue, green, pink, etc.

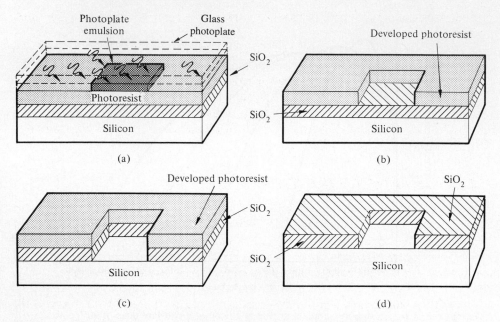

Fig. 1.7   Photolithography: (a) exposure of photoresist; (b) developed photoresist; (c) etched silicon dioxide; (d) window in the silicon dioxide.

The SiO_2 is next coated with a liquid photoresist material that is impervious to acid etching, but is sensitive to certain types of light, usually ultraviolet light. The liquid photoresist is typically applied by spinning the wafer at high speeds in order to produce a uniformly thin layer. After baking to harden, the photoresist is ready for exposure. The hardened photoresist is similar to a photographic emulsion. Figure 1.7(a) illustrates the next step, that of placing a photomask (similar to a developed picture negative) over the wafer and exposing with ultraviolet light the photoresist material in those areas where the SiO_2 is to remain. The light causes those areas of photoresist to polymerize (harden further). The photoresist is then developed much as you would an ordinary photographic film. The area not exposed to ultraviolet light is washed away, leaving the SiO_2 bare only in the unexposed areas of Fig. 1.7(a), as illustrated by Fig. 1.7(b).

A hydrofluoric-based acid that does not attack the photoresist or pure silicon is used to etch the window in the SiO_2, as shown in Fig. 1.7(c). The photoresist is then stripped from the SiO_2, leaving a window in the SiO_2 open to the silicon surface. The wafer is now ready for the thermal diffusion of impurity atoms through the window in the SiO_2, as illustrated in Fig. 1.7(d).

The basic process of photolithography is applied in many different combinations, and we have presented only one. For example, the photoresist could be of the opposite type where, with light exposure, it becomes soluble in the developer. Hence, those areas exposed to the ultraviolet light would be removed. Regardless of the specific details, a photomask, photoresist, light, and acid etch are used to open windows in specific areas of the SiO_2.

In the new era of very large-scale integrated circuits (VLSI), extending into ultra large-scale integrated circuits, called ULSI, the wavelength of light is too long for the geometries needed. Fringing becomes a problem, and x-ray or electron beam photolithography is used to expose special photoresists. For ion implantation, photolithography is used to open windows in the SiO_2, photoresist, or metals that are used to mask the surface of the silicon. The requirement is that the thickness be large enough to stop the implanted ions from reaching the silicon.

1.3   DIODE SYMBOLS AND DEFINITIONS

The standard p-n junction diode has as its symbol an arrow to indicate the direction of easy current flow. Figure 1.8 shows the symbol and the definition for positive current and voltage. With positive voltage applied to the p-region and negative voltage applied to the n-region, the diode is said to be forward biased and the current increases rapidly with small increases in voltage. This is considered the easy direction for current flow. Reverse bias occurs when the p-region is negative with respect to the n-region; that is, V_A is a negative number and very little current flows backward through the arrow of the symbol. The next two chapters are concerned with developing an explanation as to why the diode behaves nonlinearly; that is, why current flows more easily in one direction than the other. It is this ability to have nonsymmetric current flow which enables the diode to have many very important applications.

1.4   SUMMARY

This chapter has introduced how a p-n junction diode can be fabricated by diffusing, alloying, or ion implanting impurities into a substrate. The result is an abrupt transition of impurities from p-type to n-type, or vice versa. The abrupt impurity distribution can

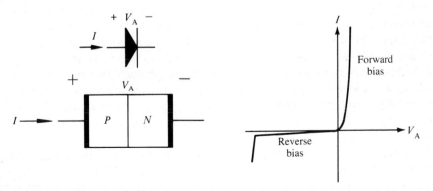

Fig. 1.8   The p-n junction diode polarity.

be approximated by the mathematical step function and is called a step junction. Alloy and epitaxially grown diodes are good examples.

Thermally diffused impurities form a graded impurity distribution that, near the metallurgical junction, can be approximated mathematically as a linear function.

Photolithography and silicon dioxide acting as a barrier to impurities create a specific area where impurities can be introduced into the substrate. The location of the metallurgical junction below the surface is a function of the dopant species, time, and temperature of the process. Impurity distributions were approximated as complementary error or Gaussian functions. An unlimited source of impurities at the surface yields a complementary error function distribution, while ion implantation and thermal diffusion from a limited source of impurities result in a Gaussian distribution. The metallurgical junction occurs at the depth where the impurity distribution is equal to the substrate doping.

PROBLEMS

1.1 The surface concentration is fixed at $N_0 = 1.8 \times 10^{+20}/cm^3$ when diffusing boron into a uniformly doped n-type substrate. The atomic diffusion constant for boron in Si at the temperature of 950 °C is $3.0 \times 10^{-15} cm^2/s$. Calculate the metallurgical junction depths for the following conditions:

(a) $N_D = 1.5 \times 10^{+16}/cm^3$; for 30 min.

(b) $N_D = 1.0 \times 10^{+15}/cm^3$; for 30 min.

(c) diffused at 950 °C for 1 hr 30 min. (i) $N_D = 1.5 \times 10^{+16}/cm^3$ (ii) $N_D = 1.0 \times 10^{+15}/cm^3$

(d) Generalize the above results into a formula.

Table P1.1 or Fig. P1.1 can be used to evaluate erfc(x).

Table P1.1

x	erfc(x)
0.70711	3.1731×10^{-1}
1.0758	1.2816×10^{-1}
1.1633	1.0000×10^{-1}
1.4142	4.5500×10^{-2}
1.8217	1.0000×10^{-2}
2.1517	2.3427×10^{-3}
2.3269	1.0000×10^{-3}
2.7511	1.0000×10^{-4}
2.7822	8.3333×10^{-5}
2.8284	6.3372×10^{-5}
3.1233	1.0000×10^{-5}
3.2120	5.5600×10^{-6}
3.2277	5.0000×10^{-6}
3.4587	1.0000×10^{-6}

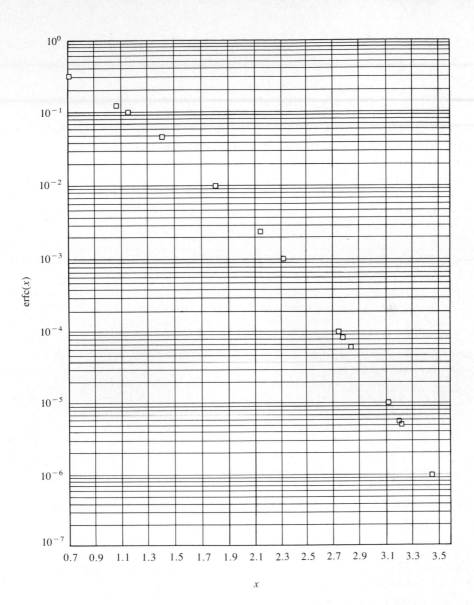

Fig. P1.1   Data from "erfc(x)"

1.2 Phosphorus is deposited on the surface of silicon at a temperature between 950 °C and 1300 °C, resulting in a nearly constant surface concentration of about $10^{+21}/\text{cm}^3$, the solid solubility of phosphorus in silicon. If the substrate is p-type and uniformly doped at $N_A = 5.0 \times 10^{+15}/\text{cm}^3$ and diffused from a constant surface concentration, calculate:

(a) the junction depth if diffused at 950 °C for (i) 30 min. and (ii) 90 min.

(b) repeat part (a) for 1050 °C

(c) the time required to place the junction at 1.2 microns with 1050 °C.

Use the formula below for obtaining the diffusion constant at various temperatures.

$$D = 3.85 \times e^{-[E_a/kT]} \quad \text{where } E_a = 3.66 \text{ eV and } T \text{ is in °K}$$

1.3 Boron atoms are diffused into silicon from a limited supply, starting with a "dose" of $Q/A = 5 \times 10^{+14}/\text{cm}^2$. After diffusing at 1050 °C ($D_{\text{boron}} = 5 \times 10^{-14} \text{ cm}^2/\text{s}$) calculate:

(a) the surface concentration after (i) 30 min. (ii) 90 min.

(b) the dopant concentration after 30 and 90 min. at (i) $x = 0.1 \ \mu\text{m}$ (ii) $x = 0.5 \ \mu\text{m}$

(c) the junction depths if the substrate is n-type doped at $N_D = 5.0 \times 10^{+15}/\text{cm}^3$ after (i) 30 min. (ii) 90 min.

1.4 The diffusion of phosphorus from a limited source with a dose of Q/A of $8 \times 10^{+14}/\text{cm}^2$ into a uniformly doped substrate doped at $3.0 \times 10^{+15}/\text{cm}^3$ is performed at several temperatures. Calculate the junction depth if

(a) diffused at 950 °C for 1 hr

(b) diffused at 1050 °C for 1 hr

Use the formula of Problem 1.2 to determine the atomic diffusion constant.

1.5 An n-type substrate uniformly doped with $N_D = 3 \times 10^{+15}/\text{cm}^3$ is thermally diffused with p-type impurities from a limited source at a temperature where $D = 5.0 \times 10^{-14} \text{ cm}^2/\text{s}$. If $Q/A = 2 \times 10^{+14}/\text{cm}^2$, calculate:

(a) the diffusion time required for $x_j = 10^{-4}$ cm. What is the surface concentration at this time?

(b) the time in part (a) for $x_j = 2 \ \mu\text{m}$. What is the surface concentration?

1.6 Boron is ion implanted into a uniformly doped n-type substrate doped at $N_D = 5.0 \times 10^{+15}/\text{cm}^3$. The implant energy is 60 KeV at a dose of $3 \times 10^{+14}/\text{cm}^2$. If the projected range is 0.20 μm and the projected straggle is 0.054 μm, calculate:

(a) the peak boron concentration

(b) the surface concentration

(c) the junction depth, x_j

1.7 To produce the "metalization pattern" in a VLSI integrated circuit, the wafer is coated with aluminum and then etched into the connecting wires. Because most of the metal is removed, the opposite-type photoresist to that of Fig. 1.7 is applied. Sketch the cross sections and top view similar to that in Fig. 1.7 for a metalization photolithography; i.e., to have a square of metal left after etching.

1.8 The doping profile of Fig. P1.8 illustrates how compensating doped material can be changed to the other type. Shown is the case where an n-type substrate uniformly doped is diffused with p-type impurities and then again with n-type to make a layer of n-p-n material. Let $N_{D2} = 3.0 \times 10^{+15}/cm^3$ and let N_A have $Q/A = 2 \times 10^{+14}/cm^2$ with $D = 5.0 \times 10^{-14}$ cm^2/s for 8 hr (total time). The surface N_{D1} has $Q/A = 8 \times 10^{+14}/cm^2$ and $D = 10^{-14}$ cm^2/s for 2 hr (total time). Assume perfect superposition of the impurities and calculate:

(a) the junction depth $x_{j\,bc}$

(b) the junction depth $x_{j\,eb}$

(c) the thickness of the p-layer

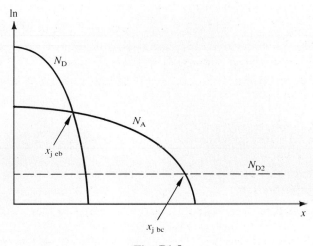

Fig. P1.8

1.9 To show the validity of the linear-graded junction approximation made from a Gaussian diffusion into a uniformly doped substrate, plot $N_A - N_D$ on a linear scale from 2.8 μm to 3.2 μm and compare it to a straight line. Use data points 0.05 μm (8 data points) apart and a least-squares analysis to the data to find the best fit to a straight line. What is the slope for the straight line? Let $N_D = 10^{+15}/cm^3$, $D = 10^{-13}$ cm^2/s, $Q/A = 10^{+14}/cm^2$, and $t = 9.064$ hr.

2 / *P-N* Junction Statics

In this chapter we focus our attention on the transition region between the p- and n-regions of the diode under static (dc) biasing conditions, examining, in turn, thermal equilibrium ($V_A = 0$), forward bias ($V_A > 0$), and reverse bias ($V_A < 0$). The region near the metallurgical junction, the transition region, is often called the *depletion region* since, as is discussed in Section 2.1, the mobile carriers in the region are reduced in number, that is, depleted in population compared to the *bulk regions* far from the junction. We first consider the junction under equilibrium conditions qualitatively to establish the source of the charge density, electric field, and potential in the depletion region. With these quantities established, the concept of the *built-in potential* (V_{bi}) is introduced. The *depletion approximation* is next invoked to solve Poisson's equation for $\mathscr{E}(x)$ and $V(x)$, assuming a step junction doping profile and $V_A = 0$. The step junction analysis is subsequently extended to forward and reverse bias. The last section outlines the analysis for the linearly graded junction and presents the results for the electric field and potential.

2.1 QUALITATIVE EQUILIBRIUM ELECTROSTATICS

Before discussing the static properties of the junction, let us clearly specify the device configuration assumed throughout most of the chapter. Figure 2.1(a) illustrates the as-fabricated abrupt p-n junction; part (b) of the figure is a one-dimensional approximation to what is clearly a three-dimensional device. The one-dimensional approximation simplifies the "equations of state" to one spatial variable and facilitates closed-form solutions of the differential equations. Such an assumption is justified if the major variables change rapidly in only one direction. Also, the assumption establishes many important concepts and in the end matches the experimental data from an actual diode exceptionally well. The entire list of major assumptions includes

1. a one-dimensional device;
2. a metallurgical junction at $x = 0$ [see Fig. 2.1(b)];
3. a step junction from N_A to N_D with uniformly doped p- and n-regions [see Fig. 2.2(a)];
4. perfect ohmic contacts far removed from the metallurgical junction.

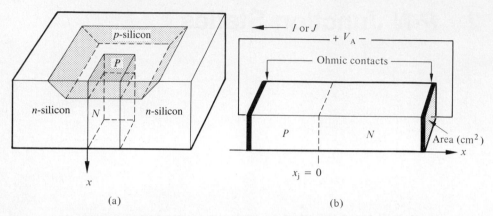

Fig. 2.1   (a) Abrupt junction *p-n* diode; (b) one-dimensional step junction.

The discussions in Volume I were principally concerned with homogeneously (uniformly) doped semiconductors. In the step junction diode, sections of the semiconductor are uniformly doped, but the doping is different on the opposite sides of the metallurgical junction. Our initial goal will be to determine and quantify the effects of this discontinuity in doping. We will begin with a qualitative examination of the structure under equilibrium conditions.

In our particular case "equilibrium" means no applied voltage ($V_A = 0$), no light shining on the device, no thermal gradients (uniform temperature), and no applied magnetic or electric fields. The situation external to the device is clearly of little interest. The internal situation, on the other hand, is quite interesting. In an attempt to ascertain the internal situation, let us first assume that *charge neutrality* exists everywhere in the device. If this were the case for the step junction of Fig. 2.2(a), the mobile carrier concentrations would be as pictured in Fig. 2.2(b) and (c). However, note that the holes on the *p*-side (p_p)* may be, for example, $10^{16}/cm^3$, while those on the *n*-side (p_n) may be $10^5/cm^3$ (if $N_A = 10^{16}/cm^3$ and $N_D = 10^{15}/cm^3$ in Si at room temperature). The holes would therefore be expected to diffuse so as to make their numbers more homogeneous throughout the material. Similar arguments hold for the electrons. Specifically, Fig. 2.2(d) and (e) indicate that the holes will diffuse from the *p*-side to the *n*-side and electrons will diffuse from the *n*-side to the *p*-side. Ideally, the diffusion process would continue until the carrier concentrations become equal on both sides of the junction. The diffusion process, however, cannot continue forever because it disrupts the charge balance between $-qN_A$ and qp_p on the *p*-side of the junction and between qN_D and $-qn_n$ on the *n*-side of the junction. When the holes diffuse away from the *p*-side, they leave be-

---

*The subscripts have been included to identify the majority and minority carriers; p_p means holes in *p*-material, p_n means holes in *n*-material, etc.

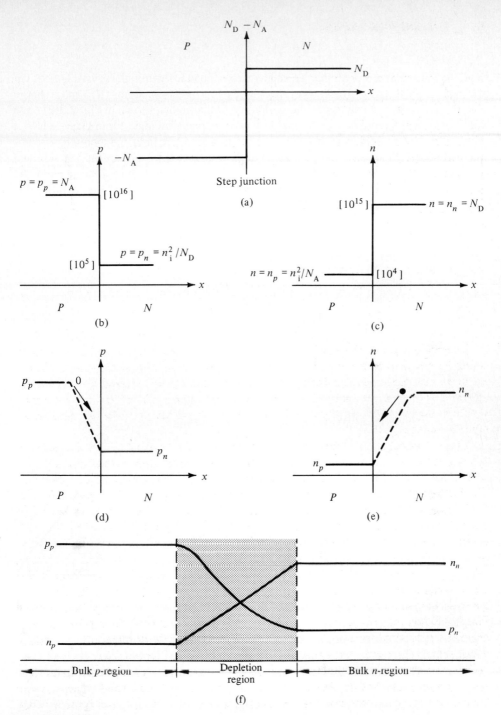

Fig. 2.2 (a) Step junction; (b) and (c) hypothetical "initial" carrier concentrations; (d) and (e) approximate equilibrium carrier concentrations; (f) combined plots. Numbers in brackets, [ ], indicate typical values. The vertical axis is a natural logarithm (ln) scale.

hind the ionized acceptor atoms (N_A^-) that are fixed in place within the crystal lattice. On the n-side the electrons diffuse away, leaving the ionized donor atoms (N_D^+) and a charge density of qN_D. Obviously, a net charge density similar to that shown in Fig. 2.3(b) must be created by the reduction of the majority carrier concentrations. From Gauss's law, a net charge density implies in turn the existence of an electric field and therefore a potential difference. Since the charge is positive on the right-hand side of the junction and negative on the left-hand side of the junction, the electric field will be directed along the negative x-axis; that is, the electric field is negative. The electric field thus opposes diffusion of holes from the p-side and electrons from the n-side. Stated another way, it acts to paste the carriers back in place, inhibiting further diffusion of the majority carriers.

The general form of the electric field can be established through the use of Eq. (2.1), where the net charge density, ρ (coulombs/cm^3), is equal to the imbalance between the charge carriers and the ions.

$$\mathcal{E}(x) = \frac{1}{K_S \varepsilon_0} \int_{-\infty}^{x} \rho(x)\, dx \qquad \text{(V/cm)} \qquad (2.1)$$

where

$$K_S = \text{relative dielectric constant of the}$$
$$\text{semiconductor (for Si } K_S = 11.8\text{),}$$
$$\varepsilon_0 = 8.854 \times 10^{-14} \qquad \text{(farad/cm),}$$

$$\rho(x) = q(p - n + N_D - N_A) \qquad \text{(coulombs/cm}^3\text{).} \qquad (2.2)$$

Performing a rough graphical integration of Fig. 2.3(b), one obtains the electric field as sketched in Fig. 2.3(c).

The V.I.P. (very important point) of this section is that the carrier concentration difference between the n- and p-regions causes the carriers to diffuse. The diffusion, however, leads to a charge imbalance. The charge imbalance in turn produces an electric field, which counteracts the diffusion so that in thermal equilibrium the *net* flow of carriers is zero. The space charge region near the metallurgical junction where the mobile carrier concentrations have been reduced below their thermal equilibrium values is called the *depletion region;* i.e., the carrier concentration of the majority has been depleted.

To continue our discussion of the qualitative aspects of the p-n junction, let us consider the potential, $V(x)$, within the depletion region and across the entire device. With a

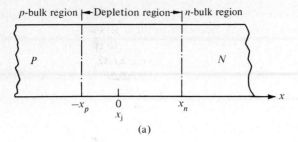

(a)

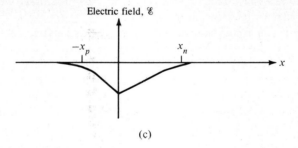

(b)

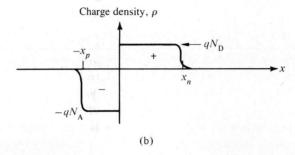

(c)

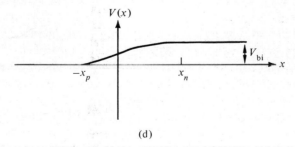

(d)

Fig. 2.3   Depletion region electrostatics.

charge density and resultant electric field present within the structure, there must also be a potential gradient. From electromagnetic field theory,

$$\mathscr{E} = -\nabla V(x) \tag{2.3a}$$

or

$$\mathscr{E} = -dV/dx, \qquad \text{for one-dimensional problems} \tag{2.3b}$$

and, integrating Eq. (2.3b),

$$V(x) = -\int_{-\infty}^{x} \mathscr{E}(x)\,dx \tag{2.4}$$

where we have arbitrarily chosen $V(-\infty) = 0$ as the reference point.* By performing a graphical integration on the electric field of Fig. 2.3(c), we obtain the potential function of Fig. 2.3(d). Note that a voltage (V_{bi}), called the *built-in potential,* exists across the depletion region of the device even at thermal equilibrium. The built-in potential can be thought of as similar to the contact potential between two dissimilar metals. In the next section we will derive a relationship for V_{bi}.

At this point the reader may well ask, "Can the energy band model be applied to the *p-n* junction, and will it yield the same information relative to ρ, $\mathscr{E}$, $V(x)$, and V_{bi} as previously discussed?" To be a consistent model, the answer to the second part of the question must be "yes." To apply the energy band model, recall that for thermal equilibrium the Fermi energy level must be a constant, independent of position. Therefore, to sketch the energy band diagram, draw a straight line for E_F on both sides of the junction. Next draw in lines for E_c, E_v, E_i parallel to E_F for the *p*-side at a large distance from the junction. Repeat this procedure for the *n*-side at a large distance from the junction. The diagram is completed by connecting the conduction and valence band edges on the two sides of the junction such that E_G is kept constant. The completed energy band model is that of Fig. 2.4.

Several facts can be deduced from the energy band diagram. Remember from Volume I that the electric field is proportional to the slope of the diagram. Hence, from the negative slope of E_c, E_v, or E_i one concludes that there is a negative electric field in the depletion region. The slope is zero at the edges of the depletion region, indi-

---

*The potential is arbitrary to within a constant and therefore we could also have selected, for example, $V(\infty) = 0$ or $V(0) = 0$.

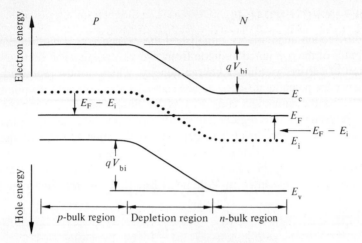

Fig. 2.4   A *p-n* junction energy band diagram at thermal equilibrium.

cating a zero electric field in the bulk regions. The maximum negative slope is near the middle, in agreement with the positioning of the maximum negative electric field, as illustrated previously in Fig. 2.3(c). Likewise, the energy difference qV_{bi} in Fig. 2.4 indicates a built-in potential difference (V_{bi}) between the ends of the device (the bulk regions). Finally, please note that the charge density can also be deduced from the energy band diagram. Since

$$\frac{d\mathscr{E}}{dx} = \frac{\rho}{K_s \varepsilon_0} \tag{2.5}$$

and

$$\mathscr{E} = \left(\frac{1}{q}\right)\left(\frac{dE_i}{dx}\right) \tag{2.6}$$

$$\rho = \left(\frac{K_s \varepsilon_0}{q}\right)\left(\frac{d^2 E_i}{dx^2}\right) \tag{2.7}$$

Thus the curvature ($d^2 E_i / dx^2$) of the energy band diagram is proportional to the charge density. Clearly, the left-hand side of the depletion region with its negative curvature implies a negative charge density, while a positive curvature signifies the existence of a positive charge density in the *n*-material. One must conclude that the energy band model can be applied to the *p-n* junction and that it provides consistent information relative to the electric field, potential, and charge density within the diode.

2.2 BUILT-IN POTENTIAL V_{bi}

The previous section qualitatively established the existence of a built-in potential (V_{bi}) across the ends of the p-n junction diode from two viewpoints. The first viewpoint was the diffusion of carriers yielding a charge density, which in turn produced an electric field and hence the potential difference. The second and equivalent viewpoint was based on the energy band diagram and constancy of E_F. Simply stated, the energy $E_c - E_F$ is different in the p and n doped regions and yields an energy difference qV_{bi}. In this section we establish an expression devoted to deriving an equation for V_{bi} that quantitatively relates V_{bi} to the doping difference between the p-region (N_A) and the n-region (N_D).

An equation for V_{bi} can be easily derived using the nondegenerate equations for the carrier concentrations presented in Volume I; Eq. (2.37) is repeated here for convenience as

$$[E_F - E_i]_{n\text{-type}} = kT \ln\left[\frac{n_n}{n_i}\right] \quad \text{on the } n\text{-side}$$

$$[E_F - E_i]_{p\text{-type}} = -kT \ln\left[\frac{p_p}{n_i}\right] \quad \text{on the } p\text{-side}$$

Inspection of Fig. 2.4 shows that the total potential difference, as obtained from E_i, is

$$[E_F - E_i]_{n\text{-type}} - [E_F - E_i]_{p\text{-side}} = kT\left[\ln\left(\frac{n_n}{n_i}\right) + \ln\left(\frac{p_p}{n_i}\right)\right] = qV_{bi}$$

$$\boxed{V_{bi} = \frac{kT}{q} \ln\left[\frac{n_n p_p}{n_i^2}\right]} \tag{2.8}$$

An equivalent derivation for V_{bi} can be presented based on thermal equilibrium ($V_A = 0$), which means that no net current flows. Stated more explicitly, $J_N = 0$, $J_P = 0$, and $J = 0$. With the net electron current zero, the current equation can be used to obtain the electric field. Setting the net electron current equal to zero yields

$$J_N = J_{N|\text{drift}} + J_{N|\text{diffusion}} = q\mu_n n\mathscr{E} + qD_N\frac{dn}{dx} = 0 \tag{2.9}$$

Note that the drift current must be equal and opposite to the diffusion current, in order for Eq. (2.9) to be zero. Solving for the electric field yields

$$\mathscr{E} = \left(\frac{-qD_N}{q\mu_n n}\right)\left(\frac{dn}{dx}\right) = -\left(\frac{D_N}{\mu_n}\right)\left(\frac{1}{n}\right)\left(\frac{dn}{dx}\right) = -\left(\frac{kT}{q}\right)\left(\frac{1}{n}\right)\left(\frac{dn}{dx}\right)$$

The last form of the equation makes use of the Einstein relationship ("*Dee* over *mu* equals *kTee* over *q*"). Using Eq. (2.4), the definition of potential, the voltage across the ends of the *p-n* junction can be written as

$$V_{\text{bi}} = -\int_{-\infty}^{\infty} \mathscr{E} \, dx = \frac{kT}{q} \int_{-\infty}^{\infty} \left(\frac{1}{n}\right)\left(\frac{dn}{dx}\right) dx = \frac{kT}{q} \int_{n(-\infty)}^{n(+\infty)} \frac{dn}{n} \qquad (2.10)$$

Integrating, we obtain

$$V_{\text{bi}} = \frac{kT}{q} \ln n \Big|_{n(-\infty)}^{n(+\infty)} \qquad (2.11)$$

Since, far from the junction on the *p*-side,

$$n_p = n(-\infty) = n_i^2/N_A \qquad (2.12a)$$

and far from the junction on the *n*-side,

$$n_n = n(+\infty) = N_D \qquad (2.12b)$$

we can write

$$V_{\text{bi}} = \frac{kT}{q}[\ln n_n - \ln n_p] = \frac{kT}{q} \ln\left[\frac{n_n}{n_p}\right] \qquad (2.13)$$

or, by substituting Eqs. (2.12a) and (2.12b) into Eq. (2.13),

$$\boxed{V_{\text{bi}} = \frac{kT}{q} \ln\left[\frac{N_D N_A}{n_i^2}\right]} \qquad (2.14)$$

Example: For silicon at room temperature ($kT = 0.026$ eV) doped $N_A = 10^{15}/\text{cm}^3$ on the *p*-side and $N_D = 10^{15}/\text{cm}^3$ on the *n*-side, where $n_i \simeq 10^{10}/\text{cm}^3$, then

$$V_{\text{bi}} = 0.026 \ln\left[\frac{10^{15}10^{15}}{10^{20}}\right] = 0.599 \text{ volts}$$

If doped at $N_A = 10^{17}/\text{cm}^3$ and $N_D = 10^{15}/\text{cm}^3$, then V_{bi} calculates to 0.718 volts. Note that the greater the doping of either side, the greater V_{bi}. The reader can establish this fact from Eq. (2.14) or the energy band diagram of Fig. 2.4. If the *p*-side doping is increased, then E_v must move closer to E_F and qV_{bi} must increase. Doping the n-side to a higher degree moves E_c closer to E_F, also increasing qV_{bi}.

Silicon has a band gap of 1.12 eV, or, in terms of kT units at room temperature, $E_G = 43.08\ kT$. To keep the semiconductor from being degenerate, E_F must be equal to or greater than 3 kT from the conduction band edge or from the valence band edge. If the step junction were doped so that E_F were at 3 kT from the band edges on each side of the device, then, with the aid of Fig. 2.4,

$$qV_{bi} = 43.08\ kT - 6\ kT = 0.9641\ \text{eV}$$

and therefore

$$V_{bi} = 0.9641\ \text{volts}$$

is the maximum value of V_{bi} at room temperature without having degenerate silicon.

SEE EXERCISE 2.1 – APPENDIX A

2.3 THE DEPLETION APPROXIMATION

The quantitative solutions for the charge density, $\mathscr{E}$, and $V(x)$ across the p-n junction under thermal equilibrium conditions are centered on the solution of Poisson's equation, repeated here for the reader's convenience:

$$\frac{d\mathscr{E}}{dx} = \frac{q}{K_S \varepsilon_0}(p - n + N_D - N_A) \tag{2.15a}$$

or

$$\frac{d^2 V}{dx^2} = -\frac{q}{K_S \varepsilon_0}(p - n + N_D - N_A) \tag{2.15b}$$

In general $\mathscr{E}$, V, p, n, N_D, and N_A are functions of x, except in the case of uniform doping, where N_D and N_A are constants. Poisson's equation in its exact form is not easily solved for most devices because p and n are in turn functions of V and $\mathscr{E}$, the unknowns. To solve* the equation in "closed form" it is necessary to make several simplifications.

---

*One can numerically solve the equation on a large digital computer by making successive guesses until an iterative solution is obtained at each point throughout the device.

A closed-form solution results in an equation for V as an explicit function of x. A particularly useful set of simplifications that allow the solution of Poisson's equation to be obtained explicitly are collectively called the *depletion approximation*. This approximation is justified if the solution can explain experimental data obtained from a real diode, which it does very well.

Figure 2.5(b) illustrates the charge density for the step junction as developed in the previous section. The depletion approximation assumes that the mobile carriers (n and p) are small in number compared to the donor and acceptor ion concentrations in the depletion region, and that the device is charge-neutral elsewhere. In the following subsection, the depletion approximation is stated mathematically and specialized to the uniformly doped p-n step junction.

2.3.1   Depletion Approximation

1. $N_A \gg n_p$ or p_p,   hence $\rho = -qN_A^-$ for $-x_p \leq x \leq 0$.
2. $N_D \gg n_n$ or p_n,   hence $\rho = qN_D^+$ for $0 \leq x \leq x_n$.
3. The charge density is zero in the bulk regions; that is, for $x > x_n$ and $x < -x_p$.

The depletion region is bounded by $-x_p$ and x_n, while the regions outside the depletion region are called the n- and p-bulk regions, respectively.

The depletion approximation, with the appropriate charge density distribution as presented in Fig. 2.5, reduces Poisson's equation (Eq. 2.15) to

$$\frac{d\mathscr{E}}{dx} = \frac{qN_D}{K_S\varepsilon_0}, \qquad \text{for } 0 \leq x \leq x_n \tag{2.16a}$$

and

$$\frac{d\mathscr{E}}{dx} = \frac{-qN_A}{K_S\varepsilon_0}, \qquad \text{for } -x_p \leq x \leq 0 \tag{2.16b}$$

where

$\mathscr{E} = 0$ in the bulk regions and at the edges of the depletion region.

Solution for $\mathscr{E}$. The electric field can be obtained for the one-dimensional step junction by simply integrating Eq. (2.16a) or (2.16b), from where one knows the electric field (at x_n and $-x_p$) to an arbitrary point inside the depletion region. Consider the p-side depletion region and Eq. (2.16b), remembering that $\mathscr{E}(-x_p) = 0$,

$$d\mathscr{E} = \frac{-qN_A}{K_S\varepsilon_0} \, dx \tag{2.17}$$

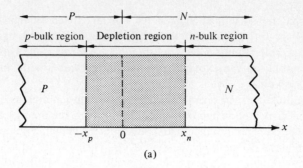

(a)

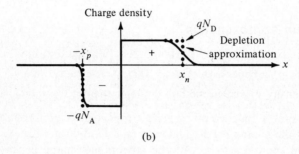

(b)

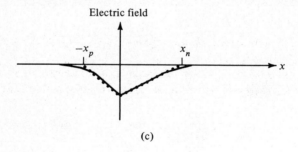

(c)

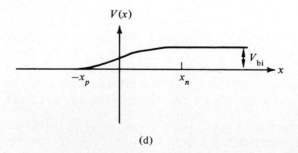

(d)

Fig. 2.5 Depletion approximation to the step junction (•••••).

$$\mathscr{E}(x) = \int_0^{\mathscr{E}(x)} d\mathscr{E} = \int_{-x_p}^{x} \left[\frac{-qN_A}{K_S\varepsilon_0} \right] dx = \frac{-qN_A}{K_S\varepsilon_0} x \Big|_{-x_p}^{x} \qquad (2.18)$$

$$\mathscr{E}(x) = \frac{-qN_A}{K_S\varepsilon_0}(x_p + x), \qquad \text{for } -x_p \leq x \leq 0 \qquad (2.19)$$

Note that the electric field is negative in agreement with our qualitative discussions. Also note that $\mathscr{E}(x)$ is the equation of a straight line with a negative slope.

The n-depletion region is obtained by integrating Eq. (2.16a) with $\mathscr{E}(x_n) = 0$. Starting at some value of x greater than zero and integrating to x_n,

$$\mathscr{E}(x) = \int_{\mathscr{E}(x)}^{0} d\mathscr{E} = \int_x^{x_n} \frac{qN_D}{K_S\varepsilon_0} dx \qquad (2.20)$$

$$-\mathscr{E}(x)\frac{qN_D}{K_S\varepsilon_0} x \Big|_x^{x_n} = \frac{qN_D}{K_S\varepsilon_0}(x_n - x) \qquad (2.21)$$

or

$$\mathscr{E}(x) = \frac{-qN_D}{K_S\varepsilon_0}(x_n - x), \qquad \text{for } 0 \leq x \leq x_n \qquad (2.22)$$

Note that Eq. (2.22) is a straight line with positive slope. Eqs. (2.19) and (2.22) are plotted in Fig. 2.5(c) as the dotted line.

As indicated in Fig. 2.5(c), the electric field must be continuous at $x = 0$ since there is no layer of sheet charge at that position. Thus from Eqs. (2.19) and (2.22), evaluated at $x = 0$,

$$\frac{-qN_A}{K_S\varepsilon_0}x_p = \frac{-qN_D}{K_S\varepsilon_0}x_n \qquad (2.23)$$

or

$$\boxed{N_A x_p = N_D x_n} \qquad (2.24)$$

Equation (2.24) multiplied by qA (where A is diode area) states that the total negative charge must equal the total positive charge. The equality is also evidenced in Fig. 2.5(b) where the area of the plot $x_p N_A$ must be equal to the area of $x_n N_D$. In Fig. 2.5 $N_A > N_D$; hence $x_p < x_n$ — an easy method of relating the doping to depletion widths, x_n and x_p.

Solution for $V(x)$. The potential function $V(x)$ within the depletion region is derived from the definition of potential; that is,

$$\frac{dV}{dx} = -\mathscr{E} \tag{2.25}$$

Combining Eq. (2.25) with the Eq. (2.19) expression for the electric field in the p-side depletion region yields

$$\frac{dV}{dx} = \frac{qN_A}{K_S\varepsilon_0}(x_p + x) \tag{2.26}$$

Separating variables and integrating gives

$$V(x) = \int_0^{V(x)} dV = \frac{qN_A}{K_S\varepsilon_0}\int_{-x_p}^x (x_p + x)\, dx \tag{2.27}$$

where we have arbitrarily selected the reference potential to be zero in the bulk p-region; that is, $V(-x_p) = 0$. We could have chosen $V(0) = 0$ or $V(x_n) = 0$. Completing the integration of Eq. (2.27) yields

$$V(x) = \frac{qN_A}{2K_S\varepsilon_0}(x_p + x)^2, \qquad \text{for } -x_p \leq x \leq 0 \tag{2.28}$$

Note that $V(x)$ is a parabola of positive curvature and is translated down the negative x-axis by x_p units.

For the n-depletion region

$$\frac{dV}{dx} = -\mathscr{E} = \frac{qN_D}{K_S\varepsilon_0}(x_p - x) \tag{2.29}$$

Since the bulk p-region was chosen to be at zero potential, the bulk n-region must be at V_{bi}; that is, $V(x_n) = V_{bi}$. Integrating,

$$\int_{V(x)}^{V_{bi}} dV = V_{bi} - V(x) = \frac{qN_D}{K_S\varepsilon_0}\int_x^{x_n} (x_n - x)\, dx$$

$$= \frac{-qN_D}{K_S\varepsilon_0}\left(x_n x - \frac{x^2}{2}\right)\Bigg|_x^{x_n} \tag{2.30}$$

$$V(x) = \frac{-qN_D}{2K_S\varepsilon_0}(x_n - x)^2 + V_{bi}, \qquad \text{for } 0 \leq x \leq x_n \tag{2.31}$$

Figure 2.5(d) illustrates the potential function $V(x)$ for the p-n step junction. By comparing Figs. 2.5(d) and 2.4, one can see that the potential function is the horizontal mirror image of the energy band diagram E_c, E_v, or E_i.

> ## SEE EXERCISE 2.2 – APPENDIX A

2.3.2   Depletion Region Width

The $\mathscr{E}(x)$ and $V(x)$ functions for the p-n junction are written in terms of the parameters N_A, N_D, x_n, and x_p. Here N_D and N_A are known from resistivity measurements. The reader should be asking the question, "How are the depletion distances x_n and x_p related to the material parameters, especially those parameters that could be measured?" To derive such a relationship, the most promising starting point is Eq. (2.31), since all the parameters except x_n are known. Remember V_{bi} was written in terms of N_A and N_D in Eq. (2.14). Since there is no dipole layer at $x = 0$, the potential function must be continuous; that is, $V(0^-) = V(0^+)$. With the aid of Eqs. (2.28) and (2.31), evaluated at $x = 0$, we conclude

$$\left[\frac{qN_A}{2K_S\varepsilon_0}\right]x_p^2 = \left[\frac{-qN_D}{2K_S\varepsilon_0}\right]x_n^2 + V_{bi} \tag{2.32}$$

Equations (2.32) and (2.24) constitute two equations and two unknowns. Solving for x_p from Eq. (2.24)

$$x_p = \left[\frac{N_D}{N_A}\right]x_n \tag{2.33}$$

and substituting into Eq. (2.32),

$$\frac{qN_A}{2K_S\varepsilon_0}\frac{N_D^2}{N_A^2}x_n^2 = \frac{qN_D^2 x_n^2}{2K_S\varepsilon_0 N_A} = \frac{-qN_D}{2K_S\varepsilon_0}x_n^2 + V_{bi} \tag{2.34}$$

We can now solve for x_n^2,

$$x_n^2 = \frac{2K_S\varepsilon_0}{q}[V_{bi}]\frac{1}{[(N_D^2/N_A) + N_D]} \tag{2.35}$$

or

$$x_n = \left[\frac{2K_S\varepsilon_0 V_{bi}}{q}\frac{N_A}{N_D(N_A + N_D)}\right]^{1/2} \tag{2.36}$$

Note that as N_D increases x_n will decrease for a fixed value of N_A. Similarly for x_p,

$$x_p = \left[\frac{2K_S\varepsilon_0 V_{bi}}{q} \frac{N_D}{N_A(N_A + N_D)} \right]^{1/2} \tag{2.37}$$

Here x_p decreases as N_A increases. The *total depletion width* (W) can be determined from x_n and x_p to be

$$W = x_n - (-x_p) = x_n + x_p \tag{2.38}$$

$$W = \left[\frac{2K_S\varepsilon_0 V_{bi}}{q(N_A + N_D)} \right]^{1/2} \left[\sqrt{\frac{N_A}{N_D}} + \sqrt{\frac{N_D}{N_A}} \right] \tag{2.39}$$

With some algebraic manipulation, Eq. (2.39) can be written in a slightly more convenient form:

$$W = \left[\frac{2K_S\varepsilon_0 V_{bi}}{q} \frac{(N_A + N_D)}{N_A N_D} \right]^{1/2} \tag{2.40}$$

Example:

$$kT = 0.026 \text{ eV}, \qquad n_i \cong 10^{10}/\text{cm}^3$$

$$N_A = 10^{16}/\text{cm}^3, \qquad N_D = 10^{15}/\text{cm}^3, \qquad V_{bi} = \frac{kT}{q} \ln \left[\frac{10^{16} 10^{15}}{(10^{10})^2} \right] = 0.659 \text{ volts},$$

$$W = \left[\frac{2 \times 11.8 \times 8.854 \times 10^{-14} \times 0.659}{1.6 \times 10^{-19}} \times \frac{(10^{16} + 10^{15})}{10^{16} \times 10^{15}} \right]^{1/2},$$

$$W = 0.9730 \times 10^{-4} \text{ cm} \quad \text{or} \quad 0.973 \text{ microns}$$

$$x_n = \left[\frac{2 \times 11.8 \times 8.854 \times 10^{-14} \times 0.659}{1.6 \times 10^{-19}} \times \frac{10^{16}}{10^{15}(10^{16} + 10^{15})} \right]^{1/2}$$

$$x_n = 0.88455 \times 10^{-4} \text{ cm} \quad \text{or} \quad 0.88455 \text{ microns}$$

Similarly from Eq. (2.37),

$$x_p = 0.08845 \times 10^{-4} \text{ cm} \quad \text{or} \quad 0.08845 \text{ microns}$$

Note that since the p-side is doped more heavily, $x_n > x_p$; that is, the junction *depletes further into the more lightly doped material*.

We remind the reader that the discussion and equations derived thus far in the chapter are for thermal equilibrium. The next section presents the case of forward bias, $V_A > 0$, and then reverse bias, $V_A < 0$.

2.4 ELECTROSTATICS OF FORWARD AND REVERSE BIAS

The electric field, charge density, and potential functions for the case of positive applied voltage ($V_A > 0$) could be developed with discussions and derivations parallel with the previous two sections. However, there is an easier approach. Consider Fig. 2.6(a), where the diode is at thermal equilibrium. No current flows and there is no voltage drop or electric field in the bulk p- and n-regions. The *ohmic* contacts of the metal-semiconductor junctions have *contact potentials* V_P and V_N that are fixed in value and depend only on the materials used to make the device. The *junction voltage* (V_j) appears across the edges of the depletion region. At thermal equilibrium $V_j = V_{bi}$. Since $V_A = 0$, writing a loop equation in Fig. 2.6(a) yields

$$V_j = V_N - 0 + V_P = V_{bi} \tag{2.41}$$

and

$$V_{bi} = V_N + V_P \tag{2.42}$$

is a fixed value depending only on the doping of the semiconductor.

The diode is forward biased when the applied voltage (V_A) has a positive potential on the p-region and a negative potential on the n-region as illustrated in Fig. 2.6(b). Since V_A is opposite in polarity relative to V_j, it must reduce the voltage across the depletion region. Writing a loop equation in Fig. 2.6(b) yields

$$V_j - V_N - V_A + V_P = V_N + V_P - V_A \tag{2.43}$$

and substituting for V_{bi} from Eq. (2.42),

$$\boxed{V_j = V_{bi} - V_A} \tag{2.44}$$

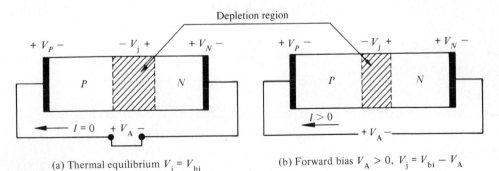

(a) Thermal equilibrium $V_j = V_{bi}$ (b) Forward bias $V_A > 0$, $V_j = V_{bi} - V_A$

Fig. 2.6 Junction potential: (a) thermal equilibrium, $V_j = V_{bi}$; (b) forward bias $V_A > 0$, $V_j = V_{bi} - V_A$.

under the assumption that little or no voltage drop occurs in the bulk p- and n-regions when $I \neq 0$.

In the previous section, for thermal equilibrium the junction potential was V_{bi} and this was used as a boundary condition in the solution of Poisson's equation. All that needs to be done to obtain a solution for the case of an applied voltage is to replace V_{bi} with $(V_{bi} - V_A)$ in Eqs. (2.36), (2.37), and (2.40).

2.4.1 The n-depletion Region, $0 \le x \le x_n$

$$x_n = \left[\frac{2K_S\varepsilon_0}{q}(V_{bi} - V_A)\frac{N_A}{N_D(N_A + N_D)} \right]^{1/2} \tag{2.45}$$

$$V(x) = (V_{bi} - V_A) - \frac{qN_D}{2K_S\varepsilon_0}(x_n - x)^2 \tag{2.46}$$

$$\mathscr{E}(x) = \frac{-qN_D}{K_S\varepsilon_0}(x_n - x) \tag{2.47}$$

2.4.2 The p-depletion Region, $-x_p \le x \le 0$

$$x_p = \left[\frac{2K_S\varepsilon_0}{q}(V_{bi} - V_A)\frac{N_D}{N_A(N_A + N_D)} \right]^{1/2} \tag{2.48}$$

$$V(x) = \frac{qN_A}{2K_S\varepsilon_0}(x_p + x)^2 \tag{2.49}$$

$$\mathscr{E}(x) = \frac{-qN_A}{K_S\varepsilon_0}(x_p + x) \tag{2.50}$$

and

$$W = \left[\frac{2K_S\varepsilon_0}{q}(V_{bi} - V_A)\left(\frac{N_A + N_D}{N_A N_D}\right) \right]^{1/2} \tag{2.51}$$

2.4.3 Forward Bias, $V_A > 0$; $V_A < V_{bi}$

Consider the results of forward bias as compared to thermal equilibrium. Since $(V_{bi} - V_A)$ is less than V_{bi} for thermal equilibrium, x_n and x_p are reduced, as illustrated in Fig. 2.7. Also shown are the effects of $V_A > 0$ on the potential, electric field, and width of the charge density; each is reduced. This may be deduced from Eqs. (2.45) through (2.51), which reduce to the thermal equilibrium relationships when $V_A = 0$. An important point about forward bias is that, for these equations to be valid,

$$V_A < V_{bi}$$

If not, Kirchhoff's voltage law is violated. The only way to remove this restriction is to allow for voltage drops in the bulk regions.

2.4.4 Reverse Bias, $V_A < 0$

The case of reverse bias requires that the applied voltage be less than zero, that is, of opposite polarity to forward bias. By inspection of Eq. (2.44), we note that the applied voltage "adds" to V_{bi} and thereby increases V_j. Since V_A is a negative number, it is therefore always less than V_{bi} — a nice result which allows the use of Eqs. (2.45) through (2.51) directly without any concern for the magnitude of V_A with respect to V_{bi}.

The increase in junction voltage, due to reverse bias, requires a larger depletion width to produce a larger fixed charge, which in turn results in a larger electric field. Figure 2.8 illustrates the effect of reverse bias on W, ρ, $\mathscr{E}$, and $V(x)$. An inspection of Eqs. (2.45), (2.48), and (2.51) also points to the increase in x_n and x_p for more negative values of V_A; that is, the depletion width W becomes wider for reverse bias. Figure 2.9 illustrates these points for a fixed value of N_A and a range of values for N_D.

SEE EXERCISE 2.3 – APPENDIX A

2.5 LINEARLY GRADED JUNCTIONS

The *ideal linearly graded* junction is an approximation to the situation where impurities are thermally diffused or ion-implanted into a semiconductor to form a p-n junction. Figure 1.5(a) illustrated the case of adding p-type impurities into an n-substrate, where near the metallurgical junction $N_A - N_D$ versus x is nearly a straight line (linear). Problem 1.8 computes this straight-line approximation.

Consider the case of p-impurities thermally diffused into an n-substrate as illustrated in Fig. 2.10(a). If the straight line has a slope of $-a$, then the impurity distribution can be described by

$$N_A(x) - N_D = -ax \tag{2.52}$$

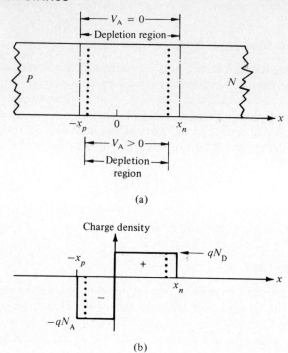

(a)

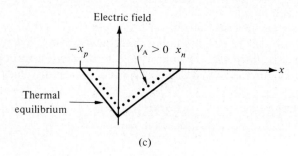

(b)

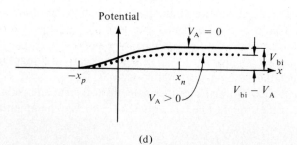

(c)

(d)

Fig. 2.7 Effect of forward bias on the diode electrostatics ($V_A > 0$, dotted lines; $V_A = 0$, unbroken lines).

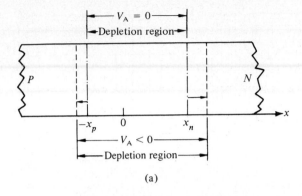

(a)

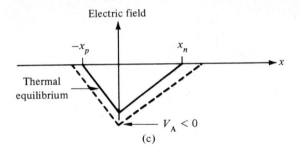

(b)

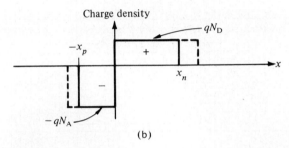

(c)

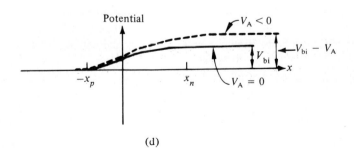

(d)

Fig. 2.8   Effect of bias on depletion region electrostatics ($V_A < 0$, dashed lines; $V_A = 0$, unbroken lines).

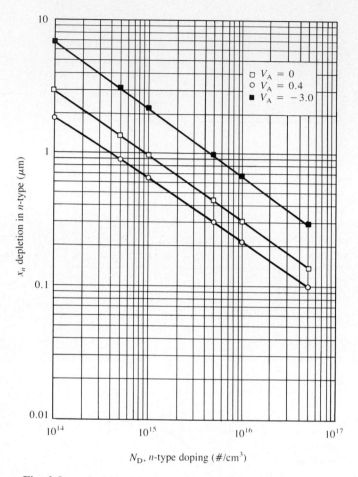

Fig. 2.9 n-depletion region for $V_A = 0$, 0.4, and -3.0 volts.

where a has the units of $\#/cm^4$ and is called the *grading constant*. Note that the metal-lurgical junction (x_j) is at $x = 0$.

The depletion approximation, as applied to the graded junction, is illustrated in Fig. 2.10(b). In particular, for the charge density,

$$\rho = qax, \quad \text{for } -x_p \leq x \leq x_n$$
$$\rho = 0, \quad \text{elsewhere}$$

(2.53)

The solution procedures for obtaining the electric field, potential, W, and V_{bi} are similar to the abrupt-junction solution and only the results are summarized here.

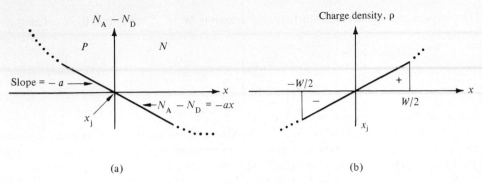

Fig. 2.10   Linearly graded p-n junction.

The symmetry of the linearly graded charge density simplifies the solution because by inspection we must have $x_n = x_p$, or the depletion width must be symmetrical about x_j. Thus

$$x_n = \frac{W}{2} \tag{2.54a}$$

and

$$x_p = \frac{W}{2} \tag{2.54b}$$

Relationships for the electrostatic variables are as follows:

$$\mathscr{E}(x) = \frac{qa}{2K_s\varepsilon_0}[x^2 - (W/2)^2], \qquad \text{for } -W/2 \leq x \leq W/2$$

$$\mathscr{E} = 0, \qquad \text{elsewhere} \tag{2.55}$$

Note that the electric field is a parabola for the graded junction as compared to a straight line for the step junction.

The potential function is as follows and is a cubic equation:

$$V(x) = \frac{qa}{6K_s\varepsilon_0}\left[2\left(\frac{W}{2}\right)^3 + 3\left(\frac{W}{2}\right)^2 x - x^3\right], \qquad \frac{-W}{2} \leq x \leq \frac{W}{2} \tag{2.56}$$

It should be no surprise that the depletion width is therefore a cube root equation.

$$W = \left[\frac{12K_s\varepsilon_0}{qa}(V_{bi} - V_A)\right]^{1/3} \tag{2.57}$$

The expression for the built-in potential requires W at $V_A = 0$, but W is calculated using V_{bi};

$$V_{bi} = \frac{2kT}{q} \ln\left[\frac{aW}{2n_i}\right], \qquad W \text{ at } V_A = 0 \qquad (2.58)$$

hence an iterative solution is required.

The case of forward and reverse bias applies to these equations in the same manner as for the step junction. Note that W increases or decreases as the one-third power for the linearly graded junction compared to the one-half power for the step junction.

2.6 SUMMARY

The step p-n junction depletion region was investigated under thermal equilibrium ($V = 0$), forward bias ($V_{bi} > 0$), and reverse bias ($V_{bi} < 0$) for the charge density (ρ), electric field ($\mathscr{E}$), and potential (V). A potential difference was produced across the depletion region (V_{bi}) as a result of the different doping and concentration on each side of the junction.

The "depletion approximation" to the diode was made so that a closed-form (equation) solution could be obtained to develop several concepts. One is that the depletion region width (W) changes with applied voltage (V_A). The second is that the depletion is further into the more lightly doped material. Another is that the maximum electric field occurs at the metallurgical junction. For a step junction, W changes as the square root, while for a linearly graded junction, W changes as the cube root of the applied voltage.

PROBLEMS

2.1 A silicon abrupt junction, approximated by a step junction, has a doping of $N_A = 5 \times 10^{+15}/\text{cm}^3$ and $N_D = 10^{+15}/\text{cm}^3$ and a cross-sectional area (A) of 10^{-4} cm^2. Assume the depletion approximation, $V_A = 0$, and $n_i = 10^{+10}/\text{cm}^3$ to:

(a) Calculate V_{bi}.

(b) Calculate x_n, x_p, and the total depletion width.

(c) What is the total positive ionic charge in the depletion width?

(d) Calculate the electric field at $x = 0$.

(e) Sketch to a relative x-axis scale the charge density and electric field.

(f) Calculate the values of n_p and p_n in the bulk regions.

(g) Draw the electron energy band diagram for the device.

(h) What percentage of W is the p-depletion region? The n-depletion region?

2.2 A silicon step junction with $N_A = 4 \times 10^{+18}/\text{cm}^3$ (assumed to be nondegenerate), $N_D = 10^{+16}/\text{cm}^3$, and $n_i = 10^{+10}/\text{cm}^3$ is at room temperature ($kT = 0.026$ eV). Calculate the following at $V_A = 0$, by assuming the depletion approximation:

(a) V_{bi}.

(b) x_n, x_p, and W.

(c) The electric field at $x = 0$.

(d) Sketch the charge density and electric field to an x-axis scale.

(e) n_p and p_n.

(f) Draw the energy band diagram for the p-n diode.

(g) What percentage of W is the p-depletion region? The n-depletion region?

2.3 A silicon p^+-n step junction diode has a doping of $N_A = 10^{+17}/\text{cm}^3$ and $N_D = 10^{+15}/\text{cm}^3$. If $kT = 0.026$ eV and $n_i = 10^{+10}/\text{cm}^3$ calculate:

(a) V_{bi}.

(b) x_n, x_p, W, and the electric field at $x = 0$.

(c) If $V_A = 0.4$ V the new values of x_n, x_p, W, and the electric field at $x = 0$.

(d) If $V_A = -3$ V the new values of x_n, x_p, W, and the electric field at $x = 0$.

(e) From the results of part (d) calculate the percentage change in x_n and W from the $V_A = 0$ case.

(f) What concept is this problem attempting to illustrate?

2.4 If a step junction silicon diode maintained at room temperature is doped such that $E_F = E_v - 2kT$ on the p-side and $E_F = E_c - E_G/4$ on the n-side, $A = 10^{-3}$ cm^2.

(a) Draw the equilibrium energy band diagram for this diode.

(b) Determine the built-in voltage, V_{bi}, as a symbolic and a numerical result. Let $kT = 0.026$ eV.

2.5 Silicon is doped as a p_1 and p_2 step junction as illustrated in Fig. P2.5, where $N_{A1} < N_{A2}$. This type of junction is called an "isotype" junction because it has the same type of doping on both sides.

(a) Sketch the energy band diagram having the more lightly doped material on the left.

(b) Derive an expression for V_{bi} from the energy band diagram.

(c) Based on the energy band diagram, sketch the approximate charge density, electric field, and potential.

(d) Explain where the charge density comes from and where it is located.

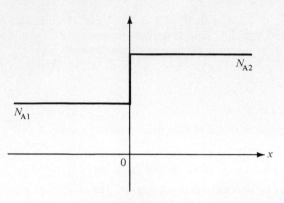

Fig. P2.5

2.6 If $N_A = 10^{+17}/\text{cm}^3$, make a table and plot x_n (in microns) vs. $\ln(N_D)$ from 10^{+14} to $10^{+17}/\text{cm}^3$ at

(a) $V_A = 0$,

(b) $V_A = 0.4$ V,

(c) $V_A = -3$. Let $n_i = 10^{+10}/\text{cm}^3$ and $kT = 0.026$ eV. Data points should be at 1, 5, and 10 for each decade. Place parts (a), (b), and (c) on the same plot.

2.7 A p^+-n step junction has $N_A = 10^{+17}/\text{cm}^3$ and $N_D = 5 \times 10^{+15}/\text{cm}^3$. If $kT = 0.026$ eV and $n_i = 10^{+10}/\text{cm}^3$:

(a) Calculate and plot x_n vs. V_A for $V_A = +0.4, -0.4, -0.8, -1, -2$, and -4 V.

(b) In the limit of large negative voltages, what is the slope of the plot?

2.8 For the linearly graded junction, assume V_{bi} is known and derive and verify the following equations:

(a) electric field, Eq. (2.55);

(b) $V(x)$, Eq. (2.56); and

(c) W, Eq. (2.57).

2.9 For the linearly graded *p-n* junction with an $a = 5 \times 10^{+19}/\text{cm}^4$, $n_i = 10^{+10}/\text{cm}^3$, and $kT = 0.026$ eV, calculate W and V_{bi} for:

(a) $V_A = 0$,

(b) $V_A = -2$,

(c) $V_A = -8$ volts.

(d) If "a" were made larger, discuss the effect on W and V_{bi} qualitatively.

3 / The Ideal Diode Volt–Ampere Characteristic

The depletion region electrostatics presented in Chapter 2 are first extended to include a visualization of carrier flow (particle flux) taking place within the device under equilibrium conditions. With the basic current-flow models firmly established, the qualitative form of the volt–ampere ($V–I$) characteristic is then deduced from a consideration of the energy band diagrams and carrier fluxes under forward and reverse biases. A first-order quantitative relationship for the $V–I$ dependence, the ideal diode equation, is next derived after establishing a "game plan" for solving the bulk p- and n-region "equations of state." In discussing the "game plan," special consideration is given the carrier densities at the edges of the depletion region. These densities are needed as boundary conditions in the analytical derivation. The analytical derivation itself yields the minority carrier concentrations, the carrier currents, and the total diode current in terms of the externally applied voltage (V_A). Much of the development in this chapter makes use of the depletion approximation and assumes a step junction device.

3.1 THERMAL EQUILIBRIUM

The diode at thermal equilibrium serves as a base upon which to build the concepts of carrier flux and potential barriers. Indeed, when forward or reverse biases are applied, the changes in the potential barrier and carrier fluxes determine the current direction and its relative magnitude.

The energy band diagram under equilibrium conditions is illustrated in Fig. 3.1. Since E_F is a constant at thermal equilibrium, the band edges (E_c and E_v) must change their position relative to E_F when making the transition from p- to n-material. As discussed previously in Chapter 2, the slope of the energy band edges is proportional to the electric field; in this case a negative slope yields a negative electric field in the depletion region. The carrier pyramids in the bulk regions of the figure crudely represent the energy distribution of the carrier densities as determined by the product of the density of states function and Fermi function. The majority and minority carriers are represented

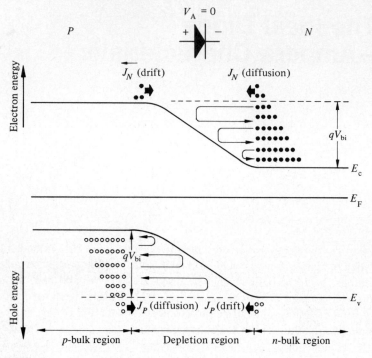

Fig. 3.1   Thermal equilibrium: energy band diagram and carrier flux.

by the relative number of appropriate symbols. It should be kept in mind that the carrier concentrations are actually decreasing exponentially with increasing energy and that the figure is only a model. See Volume I for a more detailed discussion of the carrier distributions with energy.

The reader might well ask, "Are the carriers in or near the depletion region edges drifting down the 'potential hill'?" That is, are minority electrons in the p-material drifting down the energy band diagram from left to right and holes in the n-type material drifting from right to left to seek a lower energy? This implies a current flow, which cannot be correct for thermal equilibrium. The key to resolving this quandary is that the *net current (or carrier flux) must be zero for each carrier type*. Holes cannot build up at one end of the device! Neither can electrons.

Consider the two ways in which current can flow: drift in an electric field and diffusion due to a concentration gradient. The current equations are repeated here for the reader's convenience.

$$\mathbf{J}_P = \mathbf{J}_{P|\text{drift}} + \mathbf{J}_{P|\text{diffusion}}$$ (3.1)

$$J_P = q\mu_p p \mathscr{E} - qD_P \frac{dp}{dx}$$

$$\mathbf{J}_N = \mathbf{J}_{N|\text{drift}} + \mathbf{J}_{N|\text{diffusion}} \qquad (3.2)$$

$$J_N = q\mu_n n \mathscr{E} + qD_N \frac{dn}{dx}$$

$$J = J_P + J_N \qquad (3.3)$$

For thermal equilibrium, $J = J_P = J_N = 0$. A glance at Eqs. (3.1) and (3.2) quickly reveals the need for

$$J_{N|\text{drift}} = -J_{N|\text{diffusion}}$$

and

$$J_{P|\text{drift}} = -J_{P|\text{diffusion}}$$

Consequently, in answer to our quandary, the zero net current (or carrier flux) for each carrier type under equilibrium conditions is achieved through a cancellation of the drift component by an oppositely directed diffusion component of equal magnitude.

Expanding on the foregoing conclusion, examine the carrier concentrations on the two sides of the junction. The majority carrier p_p may be, for example, $10^{16}/\text{cm}^3$, as compared to p_n, which may be $10^5/\text{cm}^3$. In going from the p-side to the n-side, the hole concentration has changed by eleven orders of magnitude! A similar case holds for electrons (see Fig. 3.2). Since, as discussed in Chapter 2, a typical depletion width may be $\sim 10^{-4}$ cm, therefore dp/dx is very large, causing the diffusion of holes from left to right in Fig. 3.1. However, the diffusing holes must climb the "potential hill," qV_{bi}, in order to enter the n-bulk region. The electric field (and hence potential) tries to "paste them back." Figure 3.1 illustrates this as the holes being reflected back to the p-region. Only those holes with energies greater than qV_{bi} can diffuse into the n-region and constitute the diffusion component of the hole current, a positive current.

Consider the minority carrier holes generated in the n-region near the depletion region edge. These holes can "fall down" the potential hill since holes like to "float" in the energy band diagram; that is, they *drift* in the electric field from right to left. This constitutes a negative hole current component.

Summarizing, at thermal equilibrium in the depletion region, the hole current components are not individually zero, that is

$$J_{P|\text{drift}} \neq 0 \quad \text{and} \quad J_{P|\text{diffusion}} \neq 0$$

but $J_P = 0$ because the components are *equal in magnitude but oppositely directed.*

Similar arguments hold true for electrons as indicated in Fig. 3.1. The carrier flux due to drift is equal and opposite to that of diffusion, yielding a *net* electron flux of zero. Thus, the net electron current is zero, with the drift and diffusion current components equal in magnitude, but oppositely directed.

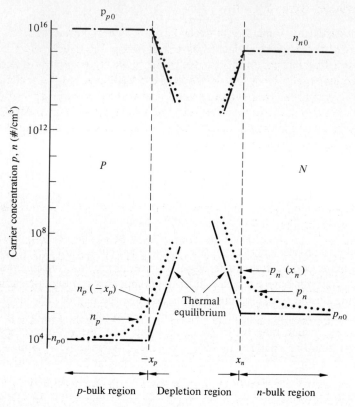

Fig. 3.2 Carrier concentrations at thermal equilibrium (–·–·–) and at forward bias (·····). The depletion region width decrease is not illustrated.

$$J_{N|\text{drift}} \neq 0 \qquad J_{N|\text{diffusion}} \neq 0 \qquad J_{N|\text{drift}} + J_{N|\text{diffusion}} = 0$$

Figure 3.1 illustrates the hole and electron particle fluxes that are equal and opposite.

Finally, in the uniformly doped bulk regions, $E_c - E_F$ is a constant and hence the electric field is zero. However, since $n = N_D = $ constant and $p = N_A = $ constant, dp/dx and dn/dx are zero. Therefore, the drift and diffusion currents (or particle fluxes) for electrons and holes are each zero and the total current is zero.

3.2 QUALITATIVE V–I CHARACTERISTICS

This section extends the carrier flux energy band diagram discussion of the previous section to describe the qualitative nature of the volt–ampere characteristic for the diode. In particular it explains why the current flow is large in one direction, but very small in the opposite direction.

3.2.1 Forward Bias, $V_A > 0$

The effect of applying a negative potential to the n-region (with respect to the p-region) on the energy band diagram is illustrated in Fig. 3.3(a). The p-side was arbitrarily chosen to remain fixed in position, with the n-side bulk region being shifted upward in electron energy by qV_A. Note that the slope of the band edges in the depletion region is reduced in magnitude as compared to what it is at thermal equilibrium. This means that the magnitude of the electric field has been reduced along with the potential difference between the ends of the diode. In fact, the energy barrier or potential hill for holes in the p-region has been reduced from qV_{bi} to

$$q(V_{bi} - V_A) \tag{3.4}$$

A similar reduction of the barrier height for the majority carrier electrons on the n-side is also illustrated in Fig. 3.3(a).

The reduced barrier for the diffusion of holes from the p-side to the n-side, for roughly the same carrier concentration gradient, yields an *increase in the hole diffusion current component* over the thermal equilibrium value. A large number of holes (p_p) have energies greater than the barrier height, $q(V_{bi} - V_A)$, as illustrated in Fig. 3.3(a), and therefore more holes can diffuse into the n-material, resulting in a larger hole diffusion current component as depicted in Fig. 3.3(b). The hole drift current component remains the same as the thermal equilibrium value (small), since a change in barrier height has no effect on the number of holes (p_n) or their ability to drift down the potential hill.

The *net* flux of holes from left to right in the diagram represents a positive hole current since the holes able to *diffuse* are more numerous than those drifting back across the junction. Holes that are able to diffuse from the p-region into the n-region are called *injected holes* once they reach the n-region bulk, and are given the name "injected minority carriers" (more about this later).

Arguments for the electrons are similar to those for the holes. The barrier for electron diffusion from the n-region into the p-region is reduced, as illustrated in Fig. 3.3(a), and a large number of electrons are able to diffuse across the junction, causing a positive current. The majority carrier electrons reaching the p-material are called "injected minority carrier electrons" upon entering the p-bulk region.

Electrons from the p-region drift down the potential hill, but this component of current remains equal to its (small) thermal equilibrium value since the carrier supply is limited by thermal generation and not by the size of the potential hill.

The net effect of forward bias is a large increase in the diffusion current components while the drift components remain fixed near their thermal equilibrium values. Since the Fermi function distributes the carriers nearly exponentially with increasing energy, one should expect the number of carriers able to diffuse to increase exponentially with the reduction of the potential barriers. This being the case, the net forward bias current should increase exponentially with V_A, as will be shown later in this chapter.

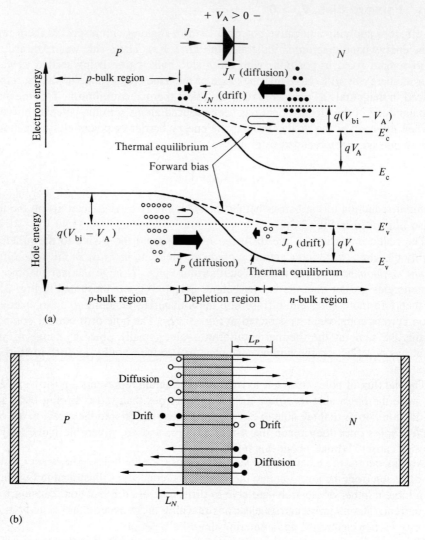

Fig. 3.3 (a) Energy band diagram for forward bias (–––––) and at thermal equilibrium (———); (b) carrier flux at forward bias, $V_A > 0$.

3.2.2 Reverse Bias, $V_A < 0$

Reverse bias has $V_A < 0$; that is, V_A is a negative number and the polarity of the applied voltage is "positive" on the n-region contact. Figure 3.4(a) illustrates the effect on the energy band diagram as compared to thermal equilibrium. The n-bulk region* is

*This choice of the n-bulk region is arbitrary and the p-bulk could have been shifted upward by $(-qV_A)$, with the n-region fixed at the thermal equilibrium level.

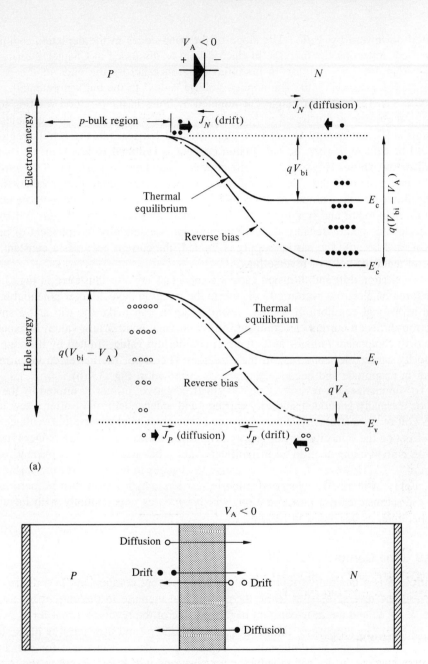

(a)

(b)

Fig. 3.4 (a) Energy band diagram for reverse bias (– · – · –) and at thermal equilibrium (————); (b) $V_A < 0$ carrier flux.

shifted downward by $-qV_A$. The slope of the band edges in the depletion region has increased, reflecting the increase in electric field as discussed in Chapter 2 for reverse bias. Also note the increase in potential difference across the ends of the device, $q(V_{bi} - V_A)$, when $V_A < 0$. The applied voltage "adds" to the built-in potential.

The increase in barrier height encountered by holes in the p-region wanting to diffuse to the n-side is illustrated in Fig. 3.4(a). From the reduced number of holes (p_p) having sufficient energy to get over the potential barrier into the n-bulk region, it should be evident that the **hole diffusion current is reduced to less than its thermal equilibrium value.** However, the drift current component of holes from the n-region to the p-region down the potential hill remains at its thermal equilibrium value (small), being limited by the number of holes thermally generated. Clearly the *net* hole current is from right to left and small in value. It is a small negative current. Figure 3.4(b) also illustrates this point. Because the drift component is essentially independent of barrier height, and because the carrier supply is limited, the current becomes a constant after several tenths of volts of reverse bias.

The electron drift and diffusion current components are also illustrated in Fig. 3.4(a). The n-region electrons wanting to diffuse to the p-region have a larger potential barrier than at thermal equilibrium. Hence, fewer electrons can make the trip and a smaller electron diffusion current, as compared to the thermal equilibrium value, is expected. The drift component remains at its thermal equilibrium value, limited by the supply of minority carrier electrons (n_p) in the p-region. Therefore, the *net* electron current is small in magnitude and negatively directed, as shown in Fig. 3.4(b).

To summarize, the reverse current is small, negative, primarily limited by the supply of thermally generated minority carriers, and independent of V_A after a few tenths of a volt of reverse bias. Further increases in the amount of reverse bias voltage have no effect on the minority carrier supply. The reverse bias voltage also reduces the diffusion current component to an insignificant value. The net result is a constant reverse current ($-I_0$). However, I_0 is very sensitive to changes in temperature, increasing with the supply of thermally generated minority carriers which are in turn proportional to n_i^2, the intrinsic carrier concentration, which increases exponentially with increasing temperature, as discussed in Volume I.

3.2.3 The Complete Circuit

Figure 3.5 is a qualitative representation of the current components in a diode under forward and reverse biasing. Note the exponential increase in the current for forward bias, $V_A > 0$, and the near-constant negative value of the reverse current for $V_A < 0$ in Fig. 3.5(a). The purpose of the metal contacts to the p- and n-material is illustrated in Fig. 3.5(b). The metal–semiconductor contacts provide a mechanism for exchanging majority carriers in the semiconductor for electrons which can travel in the external (metal wire) circuit. Consider, in particular, the source of holes at the metal–p-material contact. To create a hole an electron must jump into the external circuit. The hole then moves to the right toward the junction, eventually diffuses across the depletion region,

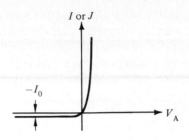

(a) Volt–ampere characteristic

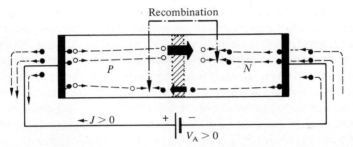

(b) Major diffusion particle fluxes for forward bias, $V_A > 0$

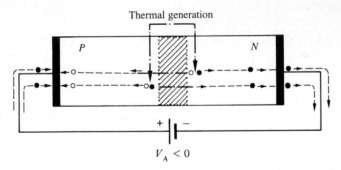

(c) Major drift particle fluxes for reverse bias, $V_A < 0$

Fig. 3.5 Diffusion and drift current components: (a) volt–ampere characteristic; (b) major diffusion particle fluxes for forward bias, $V_A > 0$; (c) major drift particle fluxes for reverse bias, $V_A < 0$.

and is injected into the *n*-region as a minority carrier. As a minority carrier the hole has a very limited lifetime and soon recombines with a majority carrier electron. This recombination in turn calls for an electron to flow from the *n*-region contact as illustrated in Fig. 3.5(b). Note the continuous flow of carriers into and out of the semiconductor via the contacts.

The n-material has majority carrier electrons that diffuse across the depletion region, are injected into the p-material, and then recombine with majority carrier holes. Replacement holes are supplied at the metal–p-semiconductor contact by electrons exiting into the external circuit, thereby generating majority carrier holes. The continuity of current is completed at the metal–n-material contact, which provides the electrons for injection.

Figure 3.5(c), appropriate for reverse biases, illustrates the thermal generation of drift current components near the depletion region edges. For example, the generation of minority carrier holes in the n-region drifts through the depletion region to the p-region, where they become majority carriers and eventually migrate to the metal–p-material contact. Electrons from the metal annihilate the excess holes, causing an external circuit electron to flow. Simultaneously, the thermally generated majority carrier electrons migrate to the metal–n-region contact and into the external circuit, completing the current loop.

The thermally generated electron–hole pairs near the p-side edge of the depletion region contribute to the reverse current by the electron's drifting across the depletion region to the n-material and as a majority carrier moves into the external circuit. At the same time, the thermally generated hole near the p-side depletion region travels to the p-region contact and is exchanged for an external circuit electron, completing the current loop.

It should be pointed out that in Fig. 3.5(b) the drift current components are not shown because they are small. The diffusion components in Fig. 3.5(c) are not shown for the same reason.

3.3 THE IDEAL DIODE EQUATION: DERIVATION GAME PLAN

This section develops a "game plan" for the quantitative solution of the basic semiconductor equations as applied to the abrupt p-n junction. The eventual goal is to derive a first-order relationship for the I versus V_A dependence of the p-n junction, known as the *ideal diode equation*. Subsequent sections carry out the mathematical details of the solution outlined in the present section.

The reader is again reminded of the explicit relationship between the p-n junction diode and the myriad of other solid state devices, and that a thorough understanding of the diode is essential to an understanding of other devices. Many of the mathematical derivations presented for the diode are used directly in modeling the bipolar, MOSFET, and junction field effect transistors. Thus, the somewhat excessive time and effort allotted to the p-n junction will be amply repaid when analyzing other devices. Also, many of the IC layout design rules for CMOS and other VLSI technologies use these concepts.

3.3.1 General Considerations

In Chapter 3 of Volume I, Section 3.4 lists the "equations of state" for all semiconductor devices: the continuity equations, Poisson's equation, and the current flow equations. These equations are to be solved in each of the three regions of the p-n junction: the

p-bulk region, the depletion region, and the n-bulk region. In Chapter 2 of Volume II we have discussed the electrostatic solution for the depletion region in detail. Several assumptions about the device must be invoked to make the $I-V_A$ solution tractable. They are justified since actual devices follow the theory over many decades of current.

1. There are no external sources of carrier generation; for example, no light.
2. The depletion approximation and the step junction are applicable.
3. The steady-state dc solution is desired; that is, all the d/dt terms in the continuity equations are zero.
4. No generation or recombination takes place in the depletion region (what goes in must come out).
5. Low-level injection is maintained in the quasi-neutral (bulk) regions of the device; this means that the number of minority carriers is always much less than the number of majority carriers in the bulk regions.
6. The electric field for the *minority carriers* is zero in the bulk regions.
7. The bulk regions are uniformly doped; that is, N_A and N_D are constants.

   With these assumptions the equations of state for the bulk n- and p-regions reduce to the following minority carrier equations.

n-type semiconductor:

$$0 = D_P \frac{d^2 \Delta p_n}{dx^2} - \frac{\Delta p_n}{\tau_p} \tag{3.5}$$

$$J_P \cong -q D_P \frac{d \Delta p_n}{dx} \tag{3.6}$$

$$p_n = p_{n0} + \Delta p_n(x) \tag{3.7}$$

p-type semiconductor:

$$0 = D_N \frac{d^2 \Delta n_p}{dx^2} - \frac{\Delta n_p}{\tau_n} \tag{3.8}$$

$$J_N \cong q D_N \frac{d \Delta n_p}{dx} \tag{3.9}$$

$$n_p = n_{p0} + \Delta n_p(x) \tag{3.10}$$

The reader should consult Section 3.4, Volume I, for the details concerning the origin of Eqs. (3.5) to (3.10).

   The plan of attack is to first solve Eq. (3.5) for $\Delta p_n(x)$ in the n-bulk region. Because Eq. (3.5) is a second-order linear differential equation, two boundary conditions are needed, one at each end of the region. We will return to a consideration of the boundary conditions later. Once $\Delta p_n(x)$ is obtained, then Eq. (3.6) is used to compute $J_P(x)$.

Since the device has only two terminals, the total current through the diode must be a constant at each point:

$$J = \text{constant} = J_N(x) + J_P(x) \tag{3.11}$$

Therefore, if the minority carrier current density $J_P(x)$ is known in the n-bulk region, the majority carrier current density $J_N(x)$ is also known from Eq. (3.11) as

$$J_N(x) = J - J_P(x) \tag{3.12}$$

Arguments for the p-bulk region and the minority carrier electrons are complementary to those of the minority carrier holes in the n-bulk region. By complementary we mean the exchange of p for n and n for p. For example, the complement of Eq. (3.5) is Eq. (3.8). A solution of Eq. (3.8) yields $\Delta n_p(x)$ and the use of Eq. (3.9) results in $J_N(x)$, the minority carrier current in the bulk p-region. The majority carrier current is obtained from Eq. (3.11) as

$$J_P(x) = J - J_N(x) \tag{3.13}$$

The perceptive student might ask, "You have outlined a plan of attack for the bulk regions, but what about the current in the depletion region?" The depletion region was assumed to have no generation or recombination. Therefore, the current through it is a constant—"what goes in must come out." No carriers are added to or subtracted from the carrier flux. Figure 3.6 illustrates this point. Note that if the **minority carrier diffusion current is known at the *edges* of the depletion region, then it is known throughout the depletion region,** and hence at the other edge of the depletion region. Specifically,

$$J_{P|\text{depl}} = J_P(x_n) = -qD_P\frac{dp_n}{dx}\bigg|_{x=x_n} \tag{3.14}$$

$$J_{N|\text{depl}} = J_N(-x_p) = qD_N\frac{dn_p}{dx}\bigg|_{x=-x_p} \tag{3.15}$$

Finally, as illustrated in Fig. 3.6, the total current is simply the sum of Eq. (3.14) and Eq. (3.15).

$$\boxed{J = J_P(x_n) + J_N(-x_p)} \tag{3.16}$$

The reader may well ask, "Where does the applied voltage (V_A) get into the act? You started out to derive a J versus V_A relationship, but V_A has yet to be even mentioned!" As it turns out, V_A enters via the boundary conditions on $p_n(x)$ and $n_p(x)$ at the edges of the depletion region in the solution to the minority carrier diffusion equations.

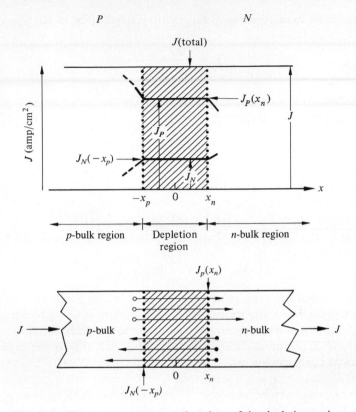

Fig. 3.6 Current components at the edges of the depletion region.

3.3.2 Boundary Conditions at x_n and $-x_p$

In Chapter 2 when V_A was applied across the terminals of the diode, the junction potential (V_j) was equated to $V_{bi} - V_A$ under the assumption that the electric field was essentially zero in the bulk regions. The assumption of "low-level injection" suggests that in the depletion region the additional currents due to the applied voltage are also small. *In the depletion region $\mathscr{E} \neq 0$ and the electron current is the difference between the* large current components, $J_{N|\text{drift}}$ and $J_{N|\text{diff.}}$. We assume that $\mathscr{E}$ and n have not changed much under low-level injection; that is,

$$J_N = q\mu_n n\mathscr{E} + qD_N\frac{dn}{dx} \cong 0 \tag{3.17}$$

and solving for the electric field yields Eq. (3.18),

$$\mathscr{E} = \frac{-qD_N\,dn/dx}{q\mu_n n} = \frac{-D_N}{\mu_n}\frac{dn/dx}{n} \tag{3.18}$$

Applying the Einstein relationship to Eq. (3.18) results in Eq. (3.19),

$$\mathscr{E} = \frac{-kT}{q} \frac{dn/dx}{n} \tag{3.19}$$

From the definition of potential, the junction voltage can be written as

$$V_j = V_{bi} - V_A = -\int_{-x_p}^{x_n} \mathscr{E}\, dx \tag{3.20}$$

Substituting Eq. (3.19) into Eq. (3.20) yields

$$V_j = V_{bi} - V_A = -\int_{-x_p}^{x_n} -\frac{kT}{q} \frac{dn/dx}{n}\, dx = \frac{kT}{q} \ln n \Big|_{n(-x_p)}^{n(x_n)} \tag{3.21}$$

$$V_{bi} - V_A = \frac{kT}{q} \ln \frac{n_n(x_n)}{n_p(-x_p)} \tag{3.22}$$

Here we have added the extra subscript to help the reader keep track of which bulk region is being discussed. Remember that for the carriers in the bulk n-region, $\Delta n_n \cong \Delta p_n \ll n_{n0}$ for the "low-level injection" requirement to be valid. Therefore $n(x_n) = n_{n0}$. Inverting Eq. (3.22) by cross multiplying and raising both sides to the exponential, we get the electron concentration ratio:

$$\frac{n_n(x_n)}{n_p(-x_p)} = e^{q(V_{bi}-V_A)/kT} = e^{[qV_{bi}/kT]}e^{[-qV_A/kT]} \tag{3.23}$$

Solving for $n(-x_p)$ results in Eq. (3.24),

$$n_p(-x_p) = n_n(x_n)e^{[-qV_{bi}/kT]}e^{[qV_A/kT]} \tag{3.24}$$

From Eq. (2.14), for thermal equilibrium,

$$V_{bi} = \frac{kT}{q} \ln\left[\frac{n_{n0}p_{p0}}{n_i^2}\right] \tag{3.25}$$

which can be inverted to obtain Eq. (3.26),

$$e^{-qV_{bi}/kT} = \frac{n_i^2}{n_{n0}p_{p0}} \tag{3.26}$$

Combining Eq. (3.24) and Eq. (3.26) yields

$$n_p(-x_p) = n_p(x_n)\frac{n_i^2}{n_{n0}p_{p0}}e^{qV_A/kT} = \frac{n_i^2}{p_{p0}}e^{qV_A/kT} \tag{3.27}$$

remembering that $n_n(x_n) \cong n_{n0}$. Since $n_i^2/p_{p0} = n_{p0}$ one concludes that

$$n_p(-x_p) = n_{p0}e^{qV_A/kT} \tag{3.28}$$

and the injected electron concentration at the boundary $-x_p$ is

$$\Delta n(-x_p) = n_{p0}(e^{qV_A/kT} - 1) \tag{3.29}$$

Note that Eqs. (3.28) and (3.29) reduce to their thermal equilibrium values when $V_A = 0$.

Complementary arguments can be used for the hole concentration at the edges of the depletion regions and result in Eqs. (3.30) and (3.31),

$$p_n(x_n) = p_{n0}e^{qV_A/kT} \tag{3.30}$$

$$\Delta p_n(x_n) = p_{n0}(e^{qV_A/kT} - 1) \tag{3.31}$$

SEE EXERCISE 3.1 – APPENDIX A

Long-Base Diode. The final boundary conditions on the excess carrier concentration in the p- and n-bulk regions are obtained by assuming that the bulk regions are very long, even infinite in length. Since the injected minority carriers have a finite lifetime (τ_p and τ_n), they cannot survive forever without recombining with a majority carrier; consequently,

$$\Delta n_p(-\infty) = 0 \tag{3.32}$$

and

$$\Delta p_n(+\infty) = 0 \tag{3.33}$$

Game Plan Summary

1. Solve the minority carrier continuity equations in the bulk regions for $\Delta p_n(x)$ and $\Delta n_p(x)$ or $p_n(x)$ and $n_p(x)$.
2. Apply two boundary conditions to each solution to determine $\Delta p_n(x)$ and $\Delta n_p(x)$ in terms of the applied voltage V_A.
3. Determine the currents $J_P(x_n)$ and $J_N(-x_p)$ from the *slope* of $\Delta n_p(x)$ and $\Delta p_n(x)$ at $-x_p$ and x_n, respectively, using Eqs. (3.14) and (3.15).

4. The total current is then the sum of the currents at the edges of the depletion region; that is,

$$J = J_P(x_n) + J_N(-x_p) \qquad (3.34)$$

3.4 THE IDEAL DIODE EQUATION: DERIVATION

To solve the minority carrier diffusion equations, Eqs. (3.5) and (3.8), we begin by selecting the special x' and x'' coordinate systems as illustrated in Fig. 3.7. The translation of the working-coordinate system simplifies the analytical form of the solutions and avoids unnecessary complications in the analysis.

3.4.1 *n*-bulk Region, $x \geq x_n$ or $x' \geq 0'$

Equation (3.35) is Eq. (3.5) rewritten in terms of the new variable x' whose origin is at $x = x_n$.

$$D_P \frac{d^2 \Delta p_n(x')}{dx'^2} - \frac{\Delta p_n(x')}{\tau_p} = 0 \qquad (3.35)$$

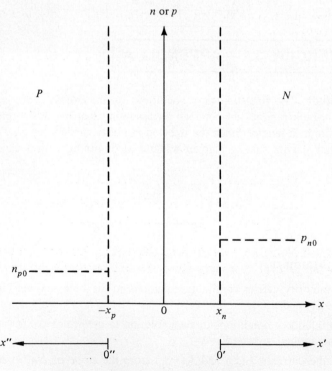

Fig. 3.7 Axis selection for bulk regions.

Dividing by D_P and taking the second term across the equal sign results in Eq. (3.36), where L_P is defined as the *minority carrier diffusion length for holes:*

$$\frac{d^2 \Delta p_n(x')}{dx'^2} = \frac{\Delta p_n(x')}{D_P \tau_p} - \frac{\Delta p_n(x')}{L_P^2} \tag{3.36}$$

$$\boxed{L_P = \sqrt{D_P \tau_p}} \quad \text{(cm)} \tag{3.37}$$

As discussed in Volume I, this is the average distance a minority carrier hole will diffuse before recombining with a majority carrier electron. Equation (3.36) is a very common differential equation and can be solved directly or by Laplace transforms. The solution is of the form

$$\boxed{\Delta p_n(x') = A_1 e^{x'/L_P} + A_2 e^{-x'/L_P}} \tag{3.38}$$

where two boundary conditions are needed to evaluate the constants A_1 and A_2. From Eq. (3.33) of the previous section, the boundary condition at infinity requires

$$\Delta p_n(\infty) = A_1 e^{\infty} + A_2 e^{-\infty} = A_1 e^{\infty} + 0 = 0 \tag{3.39}$$

Equation (3.39) can be satisfied only if $A_1 = 0$. The second boundary condition, Eq. (3.31), requires

$$\Delta p_n(x_n) = \Delta p_n(0') = p_{n0}(e^{qV_A/kT} - 1) = A_2 e^{-0'/L_P} = A_2 \tag{3.40}$$

Therefore the solution for $\Delta p_n(x)$ is

$$\Delta p_n(x') = p_{n0}(e^{qV_A/kT} - 1)e^{-x'/L_P} \tag{3.41}$$

or

$$\boxed{p_n(x') = p_{n0} + p_{n0}(e^{qV_A/kT} - 1)e^{-x'/L_P}} \tag{3.42}$$

Equation (3.42) is plotted on the right-hand side of Fig. 3.8 for $V_A > 0$ and $V_A < 0$.

Following the game plan, we next obtain the hole current by applying Eq. (3.14) to Eq. (3.41):

$$J_P(x') = -qD_P \frac{d\Delta p_n}{dx'} = -qD_P p_{n0}(e^{qV_A/kT} - 1)\left(\frac{-1}{L_P}\right)e^{-x'/L_P} \tag{3.43a}$$

$$J_P(x') = \frac{qD_P}{L_P} p_{n0}(e^{qV_A/kT} - 1)e^{-x'/L_P} \tag{3.43b}$$

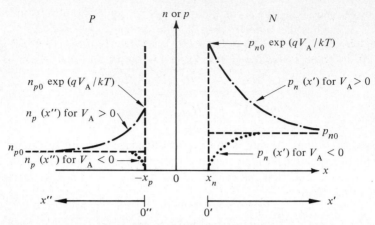

Fig. 3.8   Minority carrier concentrations in the bulk regions for $V_A > 0$ ($-\cdot-\cdot-$) and for $V_A < 0$ ($\cdots\cdots$) with $N_A > N_D$.

Evaluating at $x = x_n$ (or $x' = 0$) yields the hole current in the depletion region,

$$J_{P|\text{depl}} = J_P(x_n) = J_P(0') = q\frac{D_P}{L_P}p_{n0}(e^{qV_A/kT} - 1) \qquad (3.44)$$

Figure 3.9 illustrates the hole current at the edge and throughout the depletion and n-bulk regions.

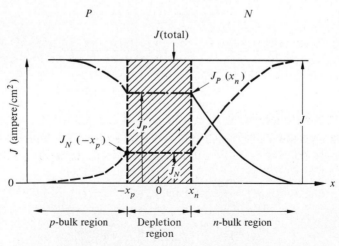

Fig. 3.9   Current density in the forward-biased case, $N_A > N_D$.

3.4.2   *p*-bulk Region, $x1 \le -x_p$ or $x'' \ge 0$

The complementary *p*-bulk solution to Eq. (3.8) can be established directly or derived step-by-step in a manner similar to the *n*-bulk derivation. Simply taking the complement of Eqs. (3.41) and (3.42) yields Eqs. (3.45) and (3.46),

$$\Delta n_p(x'') = n_{p0}(e^{qV_A/kT} - 1)e^{-x''/L_N} \tag{3.45}$$

$$n_p(x'') = n_{p0} + n_{p0}(e^{qV_A/kT} - 1)e^{-x''/L_N} \tag{3.46}$$

The *minority carrier diffusion length for electrons* has been defined as

$$L_N = \sqrt{D_N \tau_n} \quad \text{(cm)} \tag{3.47}$$

This is the average distance that the minority carrier electron will diffuse before re-combining with a majority carrier hole. The left-hand side of Fig. 3.8 illustrates these results for forward and reverse bias. By applying the complement of Eqs. (3.43) and (3.44), the electron currents $J_N(x'')$ and $J_N(0'') = J_N(-x_p)$ are readily determined to be

$$J_N(x'') = q\frac{D_N}{L_N}n_{p0}(e^{qV_A/kT} - 1)e^{-x''/L_N} \tag{3.48}$$

$$J_N(-x_p) = q\frac{D_N}{L_N}n_{p0}(e^{qV_A/kT} - 1) \tag{3.49}$$

Figure 3.9 illustrates the electron current components throughout the diode. Remember that a current directed along the x''-axis is opposite to the current flow along the *x*-axis; so a negative sign is needed in Eq. (3.15).

3.4.3   Ideal Diode Equation

The total current is obtained by adding Eq. (3.44) and Eq. (3.49) as outlined in Step 4 of the "game plan." The result is called the **ideal diode equation** or sometimes the Shockley diode equation,

$$J = q\left[\frac{D_N}{L_N}n_{p0} + \frac{D_P}{L_P}p_{n0}\right](e^{qV_A/kT} - 1) \tag{3.50}$$

Multiplying by the area of the junction (A) yields the total current

$$I = I_0(e^{qV_A/kT} - 1) \tag{3.51}$$

where the reverse saturation current has been defined as

$$I_0 = qA\left[\frac{D_N}{L_N}n_{p0} + \frac{D_P}{L_P}p_{n0}\right] \tag{3.52}$$

3.5   INTERPRETATION OF RESULTS

3.5.1   *V–I* Relationship

Further examination of the quantitative solution for the ideal diode equation leads to a deeper insight into the operation of the *p-n* junction. Figure 3.10(a) is a linear plot of Eq. (3.51) illustrating the exponential increase in current with forward bias and the small, nearly constant, reverse bias current. Examination of the exponential term in Eq. (3.51) at room temperature, where $q/kT = 38.46$ V^{-1}, indicates that a forward bias of 0.10 volts makes the exponential nearly 50 times that of the "one," and therefore, the current is exponentially dependent on the applied voltage.

$$I \cong I_0 e^{qV_A/kT}$$

If the natural logarithm is taken then

$$\ln(I) = \ln(I_0) + \frac{qV_A}{kT}$$

which is plotted as a straight line in Fig. 3.10(b). Note the similarity to the form of $y = Ax + b$ when plotted on the semilog axis.

With $V_A = -0.10$ volts, the exponential is about $1/50$ and very small compared to the "-1," leaving the current $I \cong -I_0$ independent of the reverse voltage. The reverse current no longer changes, that is, becomes saturated; hence its name "the reverse saturation current." [See Fig. 3.10(a).]

As evidenced by Eq. (3.52) and as visualized in Fig. 3.4, the reverse saturation current is determined primarily by the minority carriers n_{p0} thermally generated in the *p*-bulk region and p_{n0} generated in the *n*-bulk region. Equation (3.52) can be written in terms of the doping densities N_A and N_D:

$$I_0 = qA\left[\frac{D_N}{L_N}n_{p0} + \frac{D_P}{L_P}p_{n0}\right] = qA\left[\frac{D_N}{L_N}\frac{n_i^2}{N_A} + \frac{D_P}{L_P}\frac{n_i^2}{N_D}\right] \tag{3.53a}$$

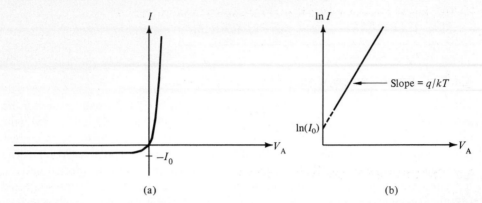

Fig. 3.10   *p-n* junction volt–ampere characteristics: (a) linear plot; (b) semilogarithmic plot.

$$I = qAn_i^2 \left[\frac{D_N}{L_N N_A} + \frac{D_P}{L_P N_D} \right] (e^{qV_A/kT} - 1) \qquad (3.53b)$$

It is important to note that n_i increases exponentially with temperature and its room-temperature value is very dependent on the band gap E_G of the material. A larger E_G value results in a smaller value of n_i. Also note that the more lightly doped side of the *p-n* junction will produce a larger number of minority carriers, and thus the larger current component. For example, a p^+-*n* junction has $N_A \gg N_D$ and, from Eq. (3.53),

$$I_0 \cong qA \left[\frac{D_P}{L_P} p_{n0} \right] = qA \left[\frac{D_P}{L_P} \frac{n_i^2}{N_D} \right] \qquad (3.54)$$

Similarly for a *p-n*$^+$ junction,

$$I_0 \cong qA \left[\frac{D_N}{L_N} n_{p0} \right] = qA \left[\frac{D_N}{L_N} \frac{n_i^2}{N_A} \right] \qquad (3.55)$$

Most diodes and many of the *p-n* junctions that occur in other devices are of the p^+-*n* or n^+-*p* type, and we shall make frequent use of these asymmetrical junctions in future analyses.

SEE EXERCISE 3.2 – APPENDIX A

3.5.2 Current Components

The forward-biased electron and hole currents illustrated in Figs. 3.8 and 3.9 are for the assumed case of $N_A > N_D$, that is, $p_{n0} > n_{p0}$. Inspection of Eqs. (3.43) and (3.48) indicates that the hole current injected into the n-region is larger than the electron current injected into the p-region when $N_A > N_D$. This is a very important point, a "V. I. P." The p^+-n junction in particular obtains most of its current from the holes injected into the n-region. Similarly, the p-n^+ junction current is composed mostly of electrons injected into the p-region.

3.5.3 Carrier Concentrations

The minority carrier concentrations are plotted for forward bias in Fig. 3.11. Cross-hatching emphasizes the injected "excess-carrier concentrations", i.e., those larger than

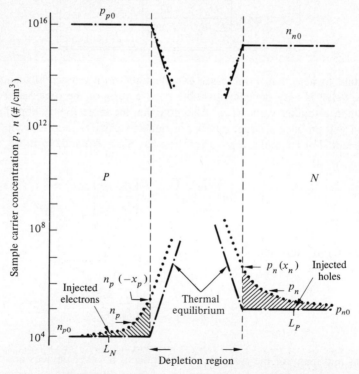

Fig. 3.11 Carrier concentrations at thermal equilibrium $(-\cdot-\cdot-)$ and at forward bias $(\cdots\cdots)$, $V_A > 0$. Depletion width decrease is not shown.

their thermal equilibrium value. From Eqs. (3.42) and (3.46), the rates of exponential decay of the injected minority carriers, as they recombine with majority carriers, are controlled by the minority carrier diffusion lengths L_P and L_N. Figure 3.11 illustrates L_P and L_N as the distance into the bulk region where the excess carriers have decreased to 37% of their value at the depletion-region edge. Since L_P and L_N are proportional to $\tau_p^{1/2}$ and $\tau_n^{1/2}$, the greater the recombination (the smaller τ_p and τ_n), the shorter the diffusion lengths. Another interpretation is that on the average a minority carrier will diffuse into the bulk region one diffusion length before recombining.

The reverse bias carrier concentrations are plotted in Fig. 3.12. In this case the "excess" carrier concentrations are negative, i.e., depleted below their thermal equilibrium values. On the average, the minority carriers thermally generated within one diffusion length of the depletion region edge are those that are encouraged to drift down the potential hill and contribute to I_0. These are the minority carriers depicted in Fig. 3.4.

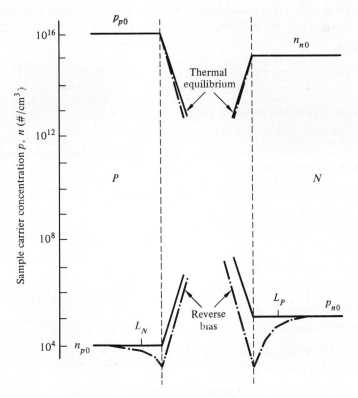

Fig. 3.12 Carrier concentrations for thermal equilibrium (———) and reverse bias (– · — · –), $V_A < 0$. Depletion width increase is not shown.

SEE EXERCISE 3.3 – APPENDIX A

3.6 SUMMARY

The thermal equilibrium, energy band diagram served as the starting point for carrier flow by diffusion and drift. The two components of hole current counteract each other for a net hole flux of zero. Similarly, the electron diffusion current was equal in magnitude but opposite in direction to the electron drift current, resulting in no net current. The drift currents were primarily controlled by the thermal generation and the carriers' flow down the potential hill. Diffusion currents are comprised of carriers with sufficient energy to surmount the potential hill and have a large concentration of carriers on one side of the depletion region compared to the other side.

The forward-biased junction permitted the diffusion currents to increase exponentially while the drift components remain fixed near their thermal equilibrium values. Holes are injected into the *n*-region and electrons are injected into the *p*-region.

PROBLEMS

3.1 Sketch the energy band diagram for a p^+-n step junction diode at:

(a) thermal equilibrium;

(b) forward bias (show with respect to thermal equilibrium);

(c) reverse bias (show with respect to thermal equilibrium);

(d) in part (a) sketch a new diagram showing the carrier flux and the four current components with respect to each other.

3.2 Starting with the minority carrier diffusion equation derive Eq. (3.45).

3.3 A silicon step junction has $N_A = 5 \times 10^{+15}/\text{cm}^3$ and $N_D = 10^{+15}/\text{cm}^3$, $D_N = 33.75$ cm^2/s, $D_P = 12.4$ cm^2/s, $n_i = 10^{+10}/\text{cm}^3$, $kT = 0.026$ eV, $A = 10^{-4}$ cm^2, $\tau_p = 0.4$ μs, and $\tau_n = 0.1$ μs. Calculate:

(a) reverse saturation current due to holes;

(b) reverse saturation current due to electrons;

(c) reverse saturation current, I_0.

(d) If $V_A = V_{bi}/2$, calculate the

  (i) hole concentration at x_n; injected hole concentration at x_n.

  (ii) hole concentration at $x' = L_p/2$.

  (iii) electron concentration and injected electron concentration at $-x_p$.

  (iv) electron concentration at $x'' = L_N/2$.

(e) If $V_A = -V_{bi}/2$, calculate the

  (i) hole concentration at x_n; injected hole concentration at x_n.

(ii) hole concentration at $x' = L_P/2$.

(iii) electron concentration and injected electron concentration at $-x_p$.

(iv) electron concentration at $x'' = L_N/2$.

(f) Calculate the total injected (or depleted) hole charge for

    (i) part (d)

    (ii) part (e)

(g) At what value of V_A and where in the diode will the assumption of "low-level injection" first be violated? Use the criterion of the minority carrier reaching 10% of the majority carrier concentration as the violation.

3.4 A p-n abrupt junction with $N_A = 10^{+17}/cm^3$ and $N_D = 5 \times 10^{+15}/cm^3$ has $\tau_p = 0.1$ μs and $\tau_n = 0.01$ μs with $kT = 0.026$ eV, $A = 10^{-4}$ cm^2, $n_i = 10^{+10}/cm^3$, $\mu_n = 801$ cm^2/V-s and $\mu_p = 438$ cm^2/V-s.

(a) Calculate the reverse saturation current due to holes.

(b) Calculate the reverse saturation current due to electrons.

(c) Calculate the total reverse saturation current.

(d) If $V_A = V_{bi}/2$, calculate the injected minority carrier currents at the edges of the depletion region. What are the injected minority carrier concentrations at 0 and 1 μm into the bulk regions?

(e) If $V_A = -V_{bi}/2$, what are the injected (depleted) minority carrier concentrations at the edges of the bulk regions? Calculate the minority carrier currents at the edges of the depletion regions.

(f) At what value of V_A and where in the diode will the assumption of "low-level injection" first be violated? Use the criterion of the minority carrier reaching 10% of the majority carrier concentration as the violation.

3.5 Two p^+-n silicon step junction diodes maintained at room temperature are identical except that $N_{D1} = 10^{+15}/cm^3$ and $N_{D2} = 10^{+16}/cm^3$. Sketch on one set of axes and compare the I-V characteristics of the two diodes.

3.6 A silicon step junction diode with a cross-sectional area $A = 10^{-4}$ cm^2 has a doping of $N_A = 10^{+17}/cm^3$ and $N_D = 10^{+15}/cm^3$. Let $\mu_n = 801$ cm^2/V-s and $\tau_n = 0.1$ μs on the p-side; and let $\mu_p = 477$ cm^2/V-s and $\tau_p = 1$ μs on the n-side.

(a) Calculate the current through the diode at room temperature $(kT/q = 0.026$ V) if

    (i) $V_A = -50$ V,

    (ii) $V_A = -0.1$ V, and

    (iii) $V_A = 0.2$ V.

(b) Assuming that the mobilities and lifetimes do not vary significantly with temperature, repeat part (a) for T = 500 °K.

(c) Summarize in your own words what has been exhibited by this problem.

3.7 If the step junction is forward biased:

(a) Derive an equation for the total injected minority carrier charge in the n-bulk region, Q_p.

(b) Take the "complement" of part (a) to obtain Q_n for the p-bulk region.

(c) If Q_p recombines every τ_p seconds, what is the hole recombination current?

(d) Repeat part (c) for Q_n which recombines every τ_n seconds.

(e) Make a statement about recombination and current as a result of parts (c) and (d), remembering that the lifetime of a minority carrier is the average time to recombine.

(f) What is the total injected recombination current (hole plus electron)? How does this equation compare with the ideal diode equation?

3.8 Figure P3.8 shows an idealization of the actual diode when reverse biased, yet it is very useful in illustrating several important concepts.

(a) Derive an expression for the number of electrons generated per second-cm^3 (generation rate) in the $n = 0$ region on the p-side at steady state.

(b) If all the electrons in the region bounded by L_N are swept out and move across the depletion region, what is the electron current flow?

(c) Repeat parts (a) and (b) for the holes in the $p = 0$ region of the n-side.

(d) What is the total minority electron and hole generation current?

(e) Compare the answer of part (d) to $-I_0$.

(f) Explain the significance and conceptual point of this problem.

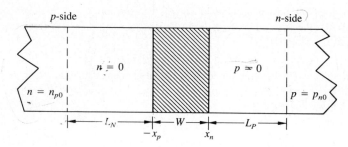

Fig. P3.8

3.9 Figure P3.9 describes what is called the "short-base diode."

(a) Derive an equation for $\Delta p_n(x')$ and sketch a plot of $p_n(x')$ if

 (i) forward biased,

 (ii) reverse biased.

(b) Derive an equation for $J_P(x')$ in the n-region and sketch a plot for forward bias.

(c) Use the previous text results for J_N in the p-region and obtain an equation for the "ideal" short-base diode I–V characteristic.

(d) Sketch plots for parts (a) and (b) for reverse bias.

(e) What is the ratio of the hole current to the total current?

(f) If X_N were to be made even smaller, expain how it will affect the injected hole current in the n-region.

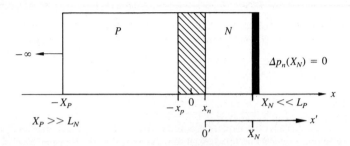

Fig. P3.9

3.10 For the "semi-short" base diode shown in Fig. P3.10:

(a) Derive an equation for $\Delta p_n(x')$ and sketch its plot. (Hint: look for hyperbolic functions.)

(b) Derive an equation for $J_P(x')$ and sketch its plot.

(c) Use the p-region expressions from the text and obtain an equation for the total current (J).

(d) Check your answer by letting $X_N \to \infty$; i.e. it should become the long-base diode.

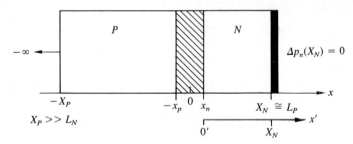

Fig. P3.10

3.11 If for the "semi-short" base diode of Fig. P3.10 the boundary condition at X_N becomes $\Delta p_n(X_N) = $ constant $= K$, derive the equation for $\Delta p_n(x')$ and sketch. Look for the solution to be in terms of hyperbolic equations.

3.12 In modern device fabrication, narrow regions in a semiconductor can be made to have almost no generation–recombination. Figure P3.12 illustrates this case for a step junction.

(a) Sketch and derive an equation for $\Delta p_n(x')$ when forward biased.

(b) Sketch and derive an equation for $J_P(x')$.

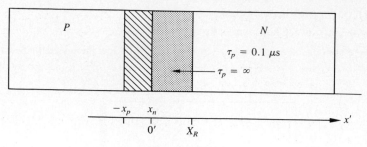

Fig. P3.12

3.13 The solar cell is a p^+-n step junction where the incident light on the device is uniformly absorbed throughout. Assume the depletion approximation still applies so that the generation due to light in the depletion region can be ignored. Given a photogeneration rate of G_L electron–hole pairs/cm^3-s, and that the p-side is heavily doped so that we need only consider the n-side and low-level injection:

(a) What is the minority carrier concentration on the n-side at a large distance ($x \to \infty$) from the metallurgical junction?

(b) Since low-level injection conditions are present, we can still use the boundary condition of

$$\Delta p_n(x_n) = p_{n0}[e^{qV_A/kT} - 1]$$

Using the boundary condition of part (a) derive an equation for the I vs. V_A of the device in steady state under illumination.

(c) Sketch the I-V characteristic on one plot with

   (i) $G_L = 0$, no light;

   (ii) $G_L = G_{L0}$;

   (iii) $G_L = 2G_{L0}$;

   (iv) $G_L = 4G_{L0}$.

3.14 A forward-biased silicon diode is sold commercially as a temperature sensor. To use it to measure temperature, it is forward biased with a constant current source and V_A is measured as a function to T as shown in Fig. P3.14.

(a) Derive an equation for $V_A(T)$ by letting D/L for holes and electrons be independent of T using the energy gap dependence of

$$E_G = 1.17 - \frac{(4.73 \times 10^{-4})T^2}{(T + 636)} \quad (eV)$$

where T is in °K.

(b) If $n_i = 1.5 \times 10^{+10}/cm^3$, $kT = 0.026$ eV, $I_0 = 10^{-15}$ A, and $I = 10^{-4}$ A at room temperature (300 °K), calculate a formula for $V_A(T)$ and plot it vs. T from 20 °C to 200 °C. Note it is nearly linear.

(c) Derive a formula for dV_A/dT (mV/°K which is the same as mV/°C) and determine the slope of the plot for the conditions of part (b).

(d) If $I = 1$ ma, repeat part (c).

(e) What is the result of parts (b), (c), and (d) telling you? How will the result affect you if you are designing a circuit?

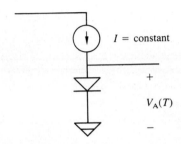

I = constant

$+$

$V_A(T)$

$-$

Fig. P3.14

4 / Deviations from the Ideal Diode

The first three chapters of this volume have developed the qualitative and quantitative descriptions of the ideal p-n junction diode. The ideal diode equation, as derived, accurately describes certain real devices over many decades of current and over a range of applied voltages. However, several conditions of applied voltage and/or temperature exist where the ideal diode fails to adequately represent physical devices. When reverse biased, the current can become more negative than $-I_0$ as a result of generation of carriers in the depletion region, and at even larger reverse voltages due to junction breakdown. Breakdown is due to one of two phenomena: avalanching or the Zener process. Forward bias deviations from the ideal occur at very small currents due to recombination in the depletion region, and at very large currents due to two effects: high-level carrier injection and ohmic voltage drops in the bulk regions and contacts.

4.1 REVERSE-BIASED DEVIATIONS FROM IDEAL

The deviations from ideal for reverse bias are illustrated in Fig. 4.1. At large values of reverse bias the magnitude of the current increases dramatically. The voltage where the current tends toward $-\infty$ is called the *breakdown voltage* (V_{BR}). Should the diode operate in this region it would not be destroyed, as would be the case for a capacitor, provided the maximum junction temperature is not exceeded. In a number of applications the diode is operated in the breakdown region and behaves like a nearly fixed voltage source over a large range of currents; hence the name "reference diode" or "regulator diode." Operated as a voltage reference in breakdown, the diode replaces a large expensive battery in many voltage regulator type circuits.

The mechanisms responsible for V_{BR} are avalanching and/or the Zener process. Avalanching is more common and will therefore be discussed first.

4.1.1 Avalanche Breakdown

As a starting point for discussing avalanche phenomena, consider the electric field in the depletion region and the thermally generated minority carriers (p_{n0} and n_{p0}) drifting

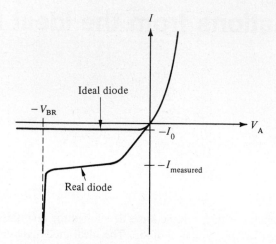

Fig. 4.1   Reverse-biased deviations from ideal.

down the potential hill, giving rise to the current $-I_0$. Chapter 2 discussed the effect of larger reverse voltages and their resultant larger electric fields in the depletion region, with the maximum electric field occurring at the metallurgical junction. Remember that the larger the electric field, the larger the drift velocity of the carriers (see the discussions on mobility in Volume I). At a critical value of electric field ($\mathscr{E}_{CR}$), the carriers, on the average, accelerate to a large enough energy that when they collide with an atom in the crystal lattice, they "free" an electron–hole pair. The colliding carrier imparts enough energy to the crystal atom that an electron in the valence band is excited to the conduction band, leaving a hole in the valence band. All three carriers are now "free" to be accelerated by the electric field, gain kinetic energy, and participate in additional carrier creating collisions; that is, an avalanche of carriers takes place. A single carrier, either an electron or a hole, creates two carriers, which in turn create additional electron–hole pairs. Figure 4.2(a) illustrates the process spatially and Fig. 4.2(b) shows the energy band.

Note that the holes entering the depletion region edge are multiplied by the avalanching phenomenon until they approach the p-bulk region and the electric field falls below the critical value. The ratio of the hole current leaving the depletion region to the hole current entering is called the multiplication factor for holes with a similar definition for electrons. The total multiplication factor is given by Eq. (4.1):

$$M = \frac{I_{\text{out}}}{I_{\text{in}}} = \frac{|I|}{I_0} \tag{4.1}$$

An expression for the multiplication factor in terms of the breakdown voltage will not be derived here, but the result is

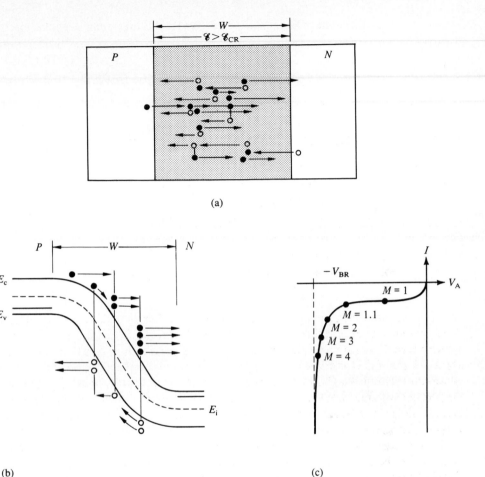

Fig. 4.2 The avalanching process: (a) spatial multiplication of electrons and holes; (b) multiplication of electrons and holes; (c) breakdown voltage vs. multiplication.

$$M = \frac{1}{1 - \left[\dfrac{|V_A|}{V_{BR}} \right]^m} \qquad (4.2)$$

where m has the range of $3 \leq m \leq 6$ depending on the semiconductor employed. Figures 4.1 and 4.2(b) illustrate that, near the knee of the V–I curve, significant multiplication occurs before the breakdown voltage is reached.

The effect of doping on the breakdown voltage will now be considered. In Chapter 2, for the step junction the maximum electric field occurred at the metallurgical junction, $x = 0$. Combining Eqs. (2.47) and (2.45), the maximum negative electric field is

$$\mathscr{E}(0) = \frac{-qN_D}{K_S\varepsilon_0}(x_n - 0) = -\left[\frac{2q}{K_S\varepsilon_0}(V_{bi} - V_A)\frac{N_A N_D}{N_A + N_D}\right]^{1/2} \tag{4.3}$$

The critical value of electric field ($\mathscr{E}_{CR}$) is a physical constant for a given semiconductor and is almost independent of doping. Replacing V_A by $-V_{BR}$ in Eq. (4.3) and squaring yields

$$\mathscr{E}_{CR}^2 = \frac{2q}{K_S\varepsilon_0}(V_{bi} + V_{BR})\frac{N_A N_D}{N_A + N_D} \tag{4.4}$$

For silicon $\mathscr{E}_{CR} \cong 2 \times 10^5$ V/cm to 8×10^5 V/cm depending upon the doping. Typically $V_{BR} \gg V_{bi}$, and one therefore concludes,

$$V_{BR} \cong \left(\frac{\mathscr{E}_{CR}^2 K_S\varepsilon_0}{2q}\right)\left[\frac{N_A + N_D}{N_A N_D}\right] \tag{4.5}$$

An examination of Eq. (4.5) indicates that any increase in doping, either n or p, results in a decrease in V_{BR}. For the p^+-n junction, where $N_A \gg N_D$, then

$$V_{BR} \propto \frac{1}{N_D} \tag{4.6}$$

and for the p-n^+ junction, where $N_D \gg N_A$, then

$$V_{BR} \propto \frac{1}{N_A} \tag{4.7}$$

The breakdown voltage in an asymmetric diode, due to avalanching, is primarily controlled by the doping of the lightly doped bulk region. Figure 4.3 illustrates the above cases more explicitly for silicon. Any V_{BR} greater than 6.5 volts is a result of avalanching.

The above discussion was based on the flat, planar p-n step junction. Diffused junctions have both vertical and lateral diffusion, which results in a nonplanar junction around the edges. As a result, the "curvature" around the edge has increased the electric fields and therefore reduced V_{BR} to below those shown in Fig. 4.3.

The temperature dependence of V_{BR} for avalanching is such that it increases as the temperature increases. This is not very evident from Eq. (4.5). It can be conceptualized as more scattering of the carriers from the lattice occurs at higher temperatures.

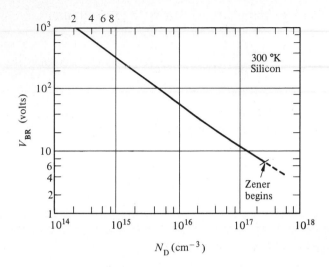

Fig. 4.3 Avalanche breakdown voltage for silicon p^+-n diode.

Therefore a larger electric field is required to accelerate the carriers to an energy high enough to cause avalanching, and it has to be done in a shorter distance before scattering and losing the energy to the lattice. The result is that a larger voltage across the device is needed for breakdown to occur.

> ### SEE EXERCISE 4.1 – APPENDIX A

4.1.2 Zener Breakdown

Zener breakdown occurs in p-n junctions that are heavily doped on both sides of the metallurgical junction. Typically Si diodes with V_{BR} of 4.5 volts or less meet the criteria for *tunneling,* another name for the Zener process. Figure 4.4 illustrates the basic idea of an electron tunneling through a potential barrier. In classical physics a particle must have an energy greater than the barrier to appear on the other side. However, quantum mechanically, if the barrier is very thin, $d < 100$ Å, the carrier may tunnel through the barrier. The two basic requirements for tunneling are:

1. A thin potential barrier; that is, the smaller d is, the larger the probability of tunneling.
2. A large number of electrons available to tunnel on one side of the barrier and a large number of empty states, at the same energy level, into which to tunnel on the other side of the barrier.

A reverse-biased p-n junction diode is illustrated in Fig. 4.5. The criteria for tunneling, as applied to the diode, are met if W is small. The depletion width is small if both

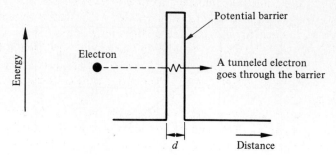

Fig. 4.4   Electron tunneling.

the p-side and the n-side have large values of N_A and N_D, respectively. With reverse bias the conduction band edge on the n-side (E_{cn}) drops below the valence-band edge on the p-side (E_{vp}), providing empty states in the conduction band of the n-material into which electrons may tunnel. Also, note that the filled electron states in the valence band on the p-side provide a larger number of electrons for tunneling. Once tunneling begins, any additional reverse bias will increase the difference $(E_{vp} - E_{cn})$ and make more electrons available to tunnel into more empty states, thereby yielding even larger reverse currents. For tunneling, V_{BR} decreases as the temperature increases.

4.1.3   Generation in W

The reverse current illustrated in Fig. 4.1 does not saturate at $-I_0$ after a few volts of reverse bias, as predicted by the ideal diode equation. In the development of the ideal

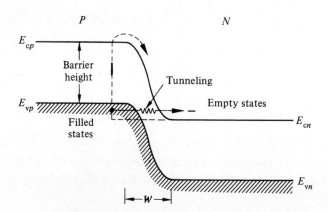

Fig. 4.5   Reverse-biased Zener breakdown.

diode equation it was assumed that the net generation–recombination rate in the depletion region was zero. When reverse biased, generation in the depletion region actually
dominates over recombination, because the carrier concentrations are less than their
thermal equilibrium values; i.e., $pn < n_i^2$. Electron–hole pairs generated in W fall down
the potential hill as illustrated in Fig. 4.6. These thermally generated carriers contribute to the current in the external circuit. Therefore a current in excess of $-I_0$ flows
through the diode.

The generation rate of electron–hole pairs (G, #/cm³-sec) can be used to calculate
the additional reverse current. The number of additional carriers created per second is
the volume times the generation rate. The additional current produced is

$$I_{R-G} = -qA \int_{-x_p}^{x_n} G \, dx \qquad (4.8)$$

The generation rate in the depletion region is approximately a constant throughout the
depletion region,

$$G = \frac{n_i}{2\tau_0} \qquad (4.9)$$

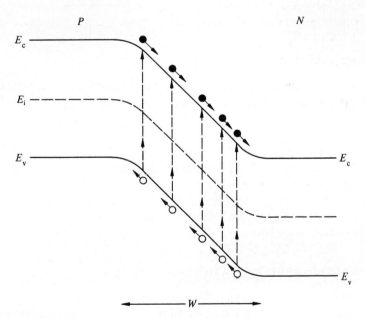

Fig. 4.6 Generation in the depletion region under reverse bias.

where τ_0 is defined as an *effective lifetime** according to Eq. (4.10),

$$\tau_0 = \frac{\tau_n + \tau_p}{2} \tag{4.10}$$

The total generation current is then obtained by substituting Eq. (4.9) into Eq. (4.8) and integrating across W:

$$I_{R-G} = -qA\frac{n_i}{2\tau_0}W \tag{4.11a}$$

Remember that $W \propto |V_A|^{1/2}$ for a step junction and therefore the generation current increases with additional reverse voltage, as depicted in Fig. 4.1. The total current is

$$I = I_0(e^{qV_A/kT} - 1) - qA\frac{n_i W}{2\tau_0} \qquad \text{(for } V_A < 0\text{)} \tag{4.11b}$$

The ideal diode equation reverse saturation current $(-I_0)$ is proportional to n_i^2, as evidenced by Eq. (3.53). The generation current is proportional to n_i and for silicon at room temperature n_i is about $10^{10}/\text{cm}^3$. A p^+-n diode has typical values of $W \cong 10^{-4}$ cm, $\tau_0 \cong \tau_p$, $L_P \cong 10^{-2}$ cm, and $N_D = 10^{14}/\text{cm}^3$. Comparing the reverse current components,

$$I_{R-G} \propto \frac{n_i}{2\tau_0}W \cong \frac{10^{10} \times 10^{-4}}{2 \times \tau_0} = \frac{5 \times 10^5}{\tau_0}$$

and

$$I_{0p} \propto \frac{n_i^2}{N_D}\frac{L_P}{\tau_p} \cong \frac{10^{20} \times 10^{-2}}{10^{14}\tau_p} \cong \frac{10^4}{\tau_p}$$

It is evident that for a silicon diode at room temperature the generation current is significantly larger as shown in Fig. 4.1. However, for a germanium diode, where n_i is about $10^{13}/\text{cm}^3$, the current terms become $(5 \times 10^8)/\tau_0$ for generation and $10^{10}/\tau_p$ for drift. A germanium diode, at room temperature under reverse bias, approximates the ideal diode equation much better than does a silicon device.

4.2   FORWARD-BIASED DEVIATIONS FROM IDEAL

The ideal diode equation, derived in Chapter 3, is plotted along with a real diode in Fig. 4.7 for forward bias voltages on a semilogarithmic scale to emphasize the large range of current values. Note that for many decades of current the ideal and the real

---

*See Volume I, Section 3.3.2.

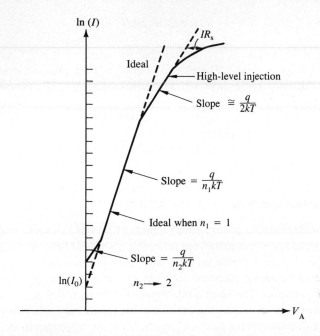

Fig. 4.7 Forward-biased deviations from the ideal.

device are really very close. The parameter "n" is called the *ideality factor* and is a measure of how close to ideal were the conditions under which the physical device was fabricated. The ideality factor is incorporated into the ideal equation by replacing q/kT by q/nkT,

$$I = I_0(e^{qV_A/nkT} - 1) \tag{4.12}$$

Under any reasonable amount of forward bias, the exponential term is much greater than -1 and

$$I \cong I_0 e^{qV_A/nkT} \tag{4.13}$$

Taking a natural logarithm yields Eq. (4.14):

$$\ln I = \ln I_0 + \frac{qV_A}{nkT} \tag{4.14}$$

When Eq. (4.14) is plotted on a semilog scale with V_A as the abscissa, we get a result similar to the xy-relationship of Eq. (4.15), where a is the $x = 0$ intercept and b is the slope:

$$y = a + bx \qquad (4.15)$$

Comparing Eq. (4.14) to Eq. (4.15), we see that the intercept is $\ln(I_0)$, as illustrated in Fig. 4.7, and the slope is q/nkT. Therefore, when $n = 1$ the physical data matches the ideal equation. For most present-day silicon devices, $1.0 \leq n \leq 1.06$ over 5 or 6 decades of current.

Three regions of nonideal behavior are illustrated in Fig. 4.7. At very small currents (near the extrapolated $V_A = 0$ intercept) the measured current is larger than the equation-predicted I. For large currents the slope decreases and eventually no specific slope can be determined. The following subsections will explain each phenomenon.

4.2.1   Recombination in *W*

Consider the nonideal region at very small values of forward current. In this region of operation the injected carrier densities are relatively small and holes leaving the p-region travel through the depletion region on their way to be injected as minority carriers into the n-region. Simultaneously, electrons are traveling from the n-region through W to be injected into the p-region. The ideal diode equation assumed that no recombination or generation occurs in the depletion region. With the carrier numbers being greater than their thermal equilibrium values $pn > n_i^2$, recombination can occur in W. Each recombination event removes an electron–hole pair. Figure 4.8 is a representation of such an event, where the recombination takes place near the center of the depletion region. Remember that for the ideal diode, the currents injected were determined by the injected carriers at the edges of the depletion region. The injected carriers at the depletion region edges were determined by the applied voltage V_A. Therefore, at a particular value of applied voltage the carriers completely traversing W are those injected as minority carriers. In the present case, more holes must enter W from the p-side because some are lost to recombination. Similarly, more electrons must enter from the n-side. The result is a larger total current than predicted by the ideal diode equation at a fixed value of V_A.

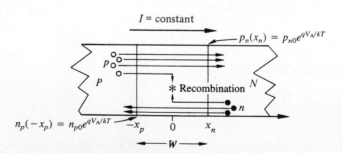

Fig. 4.8 Recombination as an additional current to the ideal diode in the depletion region under forward bias.

The total current is constant throughout the diode; therefore it is also constant in the depletion region, even though part of the hole current entering W from the p-region never reaches the n-region, and similarly, some of the electrons entering W from the n-region never reach the p-region. The electrons recombine with the holes from the p-region as illustrated in Fig. 4.8, thereby forming a **recombination current** component of the total current. The recombination current component is added to the ideal diode diffusion currents.

At very low levels of current the recombination current component dominates in silicon diodes. It should be noted that at larger current levels the recombination current is still present, but is only a small fraction of the total current. Figure 4.7 illustrates these effects for a typical silicon device. A derivation of the recombination current is possible; however, we only present the result in terms of a modified diode equation, Eq. (4.16).

$$I = I_0(e^{qV_A/n_1 kT} - 1) + q\frac{An_i}{2\tau_0}W(e^{qV_A/n_2 kT} - 1)$$ (4.16)

When the recombination current dominates at low levels of current, the second term of Eq. (4.16) is larger than the first, giving the slope of $q/n_2 kT$ as indicated in Fig. 4.7, where $n_2 > n_1 \geq 1$. Note that Eq. (4.16) is also valid for reverse bias. Typically, $n_2 \rightarrow 2$ in many devices.

4.2.2 High-Level Injection

The criterion for low-level injection is that the total minority carrier concentration always be much less than the equilibrium majority carrier concentration. For the p-n junction under forward bias, the largest minority concentration occurs at the depletion region edges; for example, see Fig. 4.9(a). At high-level injection the excess minority concen-

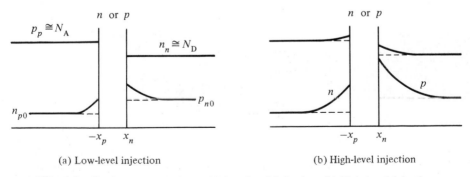

(a) Low-level injection　　　　　　　　　(b) High-level injection

Fig. 4.9　Carrier concentrations: (a) low-level injection; (b) high-level injection.

tration approaches the majority concentration. If the bulk region is to maintain charge neutrality, the majority carrier concentration has also increased significantly above its equilibrium value as illustrated in Fig. 4.9(b). The original assumption of low-level injection resulted in the definition of the recombination term to be approximated by $\Delta p/\tau_p$ and $\Delta n/\tau_n$. High-level injection requires a different recombination term with a redefinition of carrier lifetimes. It is beyond the level of this text to derive the relationship. However, the net result of high-level injection is that the ideality factor changes to about $n = 2$ at larger current levels.

SEE EXERCISE 4.2 – APPENDIX A

4.2.3   Bulk Region Effects

In the derivation of the ideal diode equation, it was assumed that the electric field in the bulk n- and p-regions was approximately zero for the minority carriers and that no voltage drop existed across the ohmic contacts. For most modern devices these are good assumptions at the lower current levels. At large current levels, however, the bulk resistance can produce a significant voltage drop and the applied voltage (V_A) is larger than the voltage across the depletion region. Also, the metal–silicon contacts can behave like small resistors, adding to the voltage drop. Usually these two effects are combined into a resistor R_s, the *series resistance*. Figure 4.7 illustrates its effect on the diode volt–ampere characteristic.

One method used to measure R_s is to plot the voltage drop from ideal (ΔV) versus current on a linear plot as outlined in Fig. 4.10. If all the data points fall on a straight

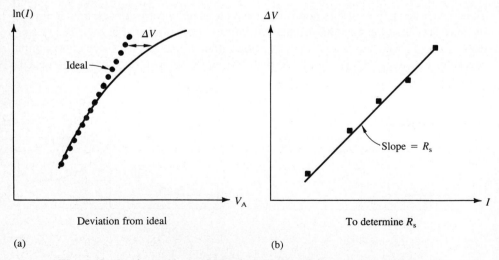

(a)   (b)

Fig. 4.10   To determine R_s obtain (a) deviation of ΔV at I, (b) plot ΔV vs. I.

line, then R_s can be determined from the slope. A word of caution is in order. In general, R_s effects can occur at lower currents than high-level injection or vice versa. The I-versus-V_A plots like those in Fig. 4.7 are routinely obtained by pulsing the current so as to avoid any self-heating effects on the results.

4.3 SUMMARY

The deviations in current and voltage of a physical diode from those of the ideal diode were considered first with reverse bias and then with forward bias. When reverse biased, the deviations were attributed to avalanche or Zener breakdown and to generation of electron–hole pairs in the depletion region. The forward-biased deviations were due to recombination in the depletion region, high-level injection, and the ohmic resistances of the bulk regions and metal contacts.

Avalanche breakdown occurred at reverse bias voltages sufficiently large that the electric field in the depletion region exceeded its critical value. At the critical value of electric field, the carriers gain enough energy so that in colliding with a crystal atom they generate an electron–hole pair. The newly generated carriers also gain energy and cause additional avalanche generation until large currents are created.

Zener breakdown is obtained from heavily doped p^+-n^+ junctions, where electrons can tunnel directly from the valence band to the conduction band. A small change in voltage results in larger numbers of carriers being able to tunnel.

Thermal generation of electrons and holes in the depletion width of the reverse-biased diode gives rise to a current component in addition to $-I_0$. The amount of generation controls the size of the current; that is, I_{R-G} is controlled by τ_0, n_i, and W.

Recombination of electrons with holes as they traverse the depletion region in a forward-biased diode produces a larger current component than that predicted by the ideal diode equation. Only at small values of total current is the recombination in the W component of significance.

When the current density is quite large, either high-level injection or the series resistance of the device, or both, becomes important. High-level injection has the effect of changing the ideality factor to about two. The series resistance increases the voltage drop across the diode for a given current.

PROBLEMS

4.1 A p^+-n silicon step junction is doped $N_A = 10^{+18}/\text{cm}^3$ and $N_D = 10^{+16}/\text{cm}^3$ where $\mathscr{E}_{CR} = 4 \times 10^{+5}$ V/cm with $n_i = 10^{+10}/\text{cm}^3$ ($kT = 0.026$ eV). Calculate:

(a) V_{BR};

(b) the depletion width at V_{BR}.

(c) If $N_D = 10^{+17}/\text{cm}^3$ repeat part (a).

4.2 Sketch the electric field in the depletion region of an avalanching step junction showing $\mathscr{E}_{CR}$ and indicate where avalanching is occurring. Repeat for a larger reverse voltage.

4.3 A silicon step junction with an area of 10^{-4} cm^2 is doped $N_A = 5 \times 10^{+15}$/cm^3 and $N_D = 10^{+15}$/cm^3. Let $n_i = 10^{+10}$/cm^3, $kT = 0.026$ eV, $D_N = 33.75$ cm^2/s, and $D_P = 12.4$ cm^2/s. If $\tau_p = 0.4$ μs and $\tau_n = 0.1$ μs, calculate:

(a) generation rate in W;

(b) generation current when $V_A = -0.1$ and -10 volts;

(c) the ratio I_{R-G}/I_0 at $V_A = -10$ volts;

(d) recombination current at $V_A = +0.1$ V and the ratio of $I_{Rec}/I_{diffusion}$;

(e) the value of V_A where half the current is recombination and half is diffusion.

4.4 The nonideal generation current in the forward-biased diode is often modeled in CAD programs in parallel. If $I_{01} = 10^{-15}$ A, $n_1 = 1.00$ and $I_{02} = 10^{-13}$ A, $n_2 = 2.00$,

(a) calculate the value of V_A where the two diode currents are equal.

(b) Plot on semilog paper the composite I-V_A from 10 fA to 1 mA.

4.5 A diode has the same specifications as those given in Problem 4.3. Plot Eq. (4.16) on semilog paper from 350 mV to 600 mV; show the recombination, diffusion, and total current components. Is the recombination plot a straight line? Explain.

4.6 Let $I_0 = 10^{-14}$ A and $n = 1.00$ for a step junction and $R_s = 20$ Ω, then calculate the current at which the applied voltage differs from the ideal by

(a) 10%.

(b) If $R_s = 2$ Ω repeat part (a).

4.7 Explain where in a p^+-n diode high-level injection will first occur and how it affects the ln[I] vs. V_A plot. What is the relationship between N_D of the device and the current at the onset of high-level injection?

4.8 A p-n diode has the measured I-V_A data of Table P4.8 and plotted in Fig. P4.8. Let $A = 10^{-4}$ cm^2, $n_i = 10^{+10}$/cm^3, and $kT = 0.026$ eV to calculate the following (Hint: use a least-squares fit to several selected data points to get the best fit):

(a) I_{01} and n_1, the diffusion current;

(b) I_{02} and n_2, the recombination current.

(c) At large currents is the deviation from the ideal resistance or high-level injection? If R_s, what is its value?

(d) Calculate the effective depletion region lifetime if $W = 0.8$ μm.

Table P4.8

V_A	I	V
0.001	5.0972E−12	1.0000E−03
0.05	1.3079E−11	5.0000E−02
0.1	2.9214E−11	1.0000E−01
0.15	8.4512E−11	1.5000E−01
0.2	2.2925E−10	2.0000E−01
0.25	6.0849E−10	2.5000E−01
0.3	1.6046E−09	3.0000E−01
0.35	4.2379E−09	3.5000E−01
0.4	1.1307E−08	4.0000E−01
0.45	3.0996E−08	4.5000E−01
0.5	9.0382E−08	5.0000E−01
0.55	2.9768E−07	5.5000E−01
0.6	1.1823E−06	6.0001E−01
0.65	5.7520E−06	6.5005E−01
0.7	3.2568E−05	7.0026E−01
0.75	2.0064E−04	7.5161E−01
0.8	1.2855E−03	8.1028E−01
0.81	1.8684E−03	8.2495E−01
0.82	2.7170E−03	8.4174E−01
0.83	3.9527E−03	8.6162E−01
0.84	5.7526E−03	8.8602E−01
0.85	8.3746E−03	9.1700E−01

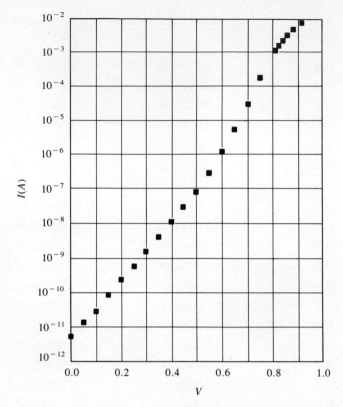

Fig. P4.8

4.9 For the diode data listed in Table P4.9, which is at large currents, determine if the deviation from ideal is caused by high-level injection or by R_s. If R_s, what is its value? Assume the ideal device has $n = 1.00$ and $I_0 = 10^{-15}$ A.

Table P4.9

I	V
5.000×10^{-4}	0.71840
1.000×10^{-3}	0.75445
5.000×10^{-3}	0.83814
1.000×10^{-2}	0.87418
5.000×10^{-2}	0.95787

4.10 (a) Derive a formula for the ΔV of the high-level injection region for Fig. P4.10. (b) If $n_1 = 1.00$ and $I_{01} = 10^{-15}$ A, calculate n_3 and I_{03} for the data of Table P4.9.

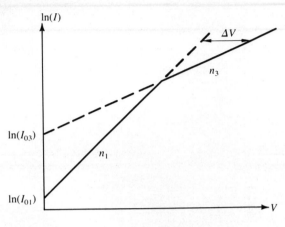

Fig. P4.10

5 / *P-N* Junction Admittance

The previous two chapters have described the response of the *p-n* junction to a dc voltage. The present chapter will investigate the response of the diode to a small-signal (sine wave) voltage superimposed upon the dc voltage. Usually the signal response is described in terms of the *small-signal admittance (Y)*, which has a real part called the *conductance* and an imaginary part called the *susceptance*. The term "small-signal" implies that the peak values of the signal current and voltage are much smaller than the dc values. Typically this means a signal voltage of several millivolts or less.

5.1 REVERSE-BIASED JUNCTION ADMITTANCE

The small-signal admittance of the reverse-biased *p-n* junction is modeled by a conductance and capacitive susceptance in parallel as illustrated in Fig. 5.1. When a small-signal, sine wave voltage v_a is applied to the diode in addition to the reverse-biased dc voltage (V_A), as illustrated in Fig. 5.2 by $v_A = v_a + V_A$, the admittance can be expressed as Eq. (5.1),

$$Y = G + j\omega C \tag{5.1}$$

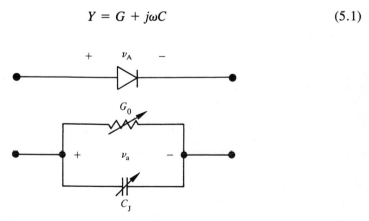

Fig. 5.1 Small-signal equivalent circuit for the reverse-biased diode; $V_A \leqq 0$.

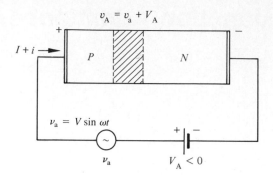

Fig. 5.2 Small-signal ac voltage plus the dc bias voltage.

For frequencies sufficiently low, where the carrier response time is much shorter than the period of the signal, Eq. (5.1) simplifies to Eq. (5.2):

$$Y = G_0 + j\omega C_J \tag{5.2}$$

Here G_0 is the low-frequency conductance which is independent of ω, but as will be derived, is dependent on the dc operating variables V_A and I. The *depletion layer capacitance* or, as often termed, *junction capacitance* C_J, is also dependent upon the dc variables and is independent of frequency. The remainder of this section is devoted to obtaining a physical insight into how the junction responds to the signal and to obtaining expressions for G_0 and C_J.

5.1.1 Reverse-Biased Depletion Capacitance

Unlike parallel metal plate capacitance, which is constant, the junction depletion capacitance changes with the applied dc (reverse) voltage as illustrated in Fig. 5.3. The capacitance decreases as V_A becomes more negative.

To explain this phenomenon, consider the depletion region at some fixed value of reverse voltage. For a step junction the depletion width (W) is determined by Eq. (2.51), repeated here for convenience and shown in Fig. 5.4(a).

$$W = \left[\frac{2K_S\varepsilon_0}{q} (V_{bi} - V_A) \frac{(N_A + N_D)}{N_A N_D} \right]^{1/2} \tag{5.3}$$

With the signal superimposed, V_A is replaced by $(V_A + v_a)$ in Eq. (5.3) and W increases or decreases by an increment ΔW. For small signals, $|v_a| \ll |V_A|$, and therefore $|\Delta W| \ll W$. Nevertheless, charge is added to or subtracted from the depletion region edges in response to v_a. When $v_a > 0$, W decreases by adding holes to the p-region and electrons to the n-region as illustrated in Fig. 5.4(b). As W becomes smaller, it must

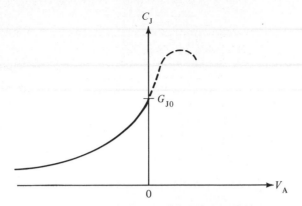

Fig. 5.3   Depletion capacitance.

do so by covering the $-qN_A$ ions with holes, the **majority carrier**, from the p-region. Similarly the $+qN_D$ charge is covered (or neutralized) with majority carrier electrons from the n-region. The net charge increments for $v_a > 0$ are illustrated in Fig. 5.4(c).

When $v_a < 0$, W must become incrementally larger by depleting majority carrier holes from the p-region and majority carrier electrons from the n-region as illustrated in Fig. 5.4(a) and (b). The charge increments for this case are illustrated in Fig. 5.4(d). Two important observations must be emphasized. First, the charge being moved is always a majority carrier charge. Majority carrier charges respond to a voltage change at roughly the dielectric relaxation time of the material. In silicon, at normal doping levels, the majority carrier response time is from about 10^{-10} to 10^{-12} sec. With such short response times, the phenomena will be independent of the frequency of v_a up to very high frequencies. The second observation is that the incremental charge diagrams of Fig. 5.4(c) and (d) are similar to the charge fluctuations on the parallel metal plates of a capacitor with area (A) and separation (W). Because of this similarity between the reverse-biased diode and the metal capacitor, the depletion (C_J) is obtained from the parallel plate capacitance formula as Eq. (5.4). Equation (5.4) is valid provided W is essentially fixed; that is, $|\Delta W|$ must be very small compared to W:

$$C_J = \frac{K_S \varepsilon_0 A}{W} \tag{5.4}$$

For small signals $|v_a| \ll |V_A|$, and when $v_A = V_A + v_a$ is substituted into Eq. 5.3 for V_A, then $|\Delta W|$ is $\ll W$. Substituting Eq. (5.3) into Eq. (5.4), remembering that $V_A < 0$, we obtain Eq. (5.5) for the depletion capacitance.

$$C_J = \frac{K_S \varepsilon_0 A}{\left[\dfrac{2K_S \varepsilon_0}{q} (V_{bi} - V_A) \dfrac{(N_A + N_D)}{N_A N_D} \right]^{1/2}} \tag{5.5}$$

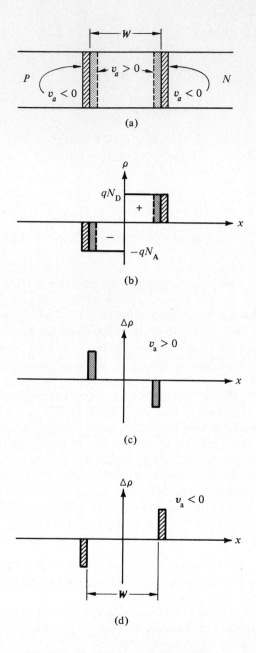

Fig. 5.4   Charge increments for C_J, the depletion capacitance.

An examination of Eq. (5.5) yields the following facts.

1. As indicated in Fig. 5.3, C_J decreases as V_A becomes more negative because W increases as discussed in Chapter 2; that is,

$$C_J \propto \frac{1}{|V_A|^{1/2}}, \qquad \text{if } |V_A| \gg V_{bi}$$

2. If N_A or N_D is increased, W decreases, thereby increasing C_J.

3. For a p^+-n junction $C_J \propto N_D^{1/2}$ on the more lightly doped side. Similarly for a p-n^+ junction $C_J \propto N_A^{1/2}$.

4. If the junction were a linearly graded junction rather than an abrupt junction, then at $|V_A| \gg V_{bi}$, W would vary as the one-third power of V_A rather than the one-half power; that is,

$$C_J \propto \frac{1}{|V_A|^{1/3}}$$

(See Chapter 2, Eq. (2.57) for further details.)

Equation (5.5) is often generalized and written in terms of the *zero-biased* depletion capacitance (C_{J0}). In Eq. (5.5) let $V_A = 0$, then the definition of C_{J0} is as follows:

$$C_{J0} = C_J|_{V_A=0} = \frac{K_S \varepsilon_0 A}{\left[\dfrac{2 K_S \varepsilon_0}{q} V_{bi} \dfrac{(N_A + N_D)}{N_A N_D} \right]^{1/2}} \tag{5.6}$$

Factoring out C_{J0} from Eq. (5.5) yields Eq. (5.7):

$$C_J = \frac{K_S \varepsilon_0 A}{\left[\dfrac{2 K_S \varepsilon_0}{q} V_{bi} \dfrac{(N_A + N_D)}{N_A N_D} \right]^{1/2} \left(1 - \dfrac{V_A}{V_{bi}} \right)^{1/2}} = \frac{C_{J0}}{\left(1 - \dfrac{V_A}{V_{bi}} \right)^{1/2}} \tag{5.7}$$

Finally, replacing the one-half power in Eq. (5.7) by m yields Eq. (5.8). The primary reason for introducing the generalized m is for handling experimental data on p-n junctions where the abruptness is usually unknown.

$$C_J = \frac{C_{J0}}{\left[1 - \dfrac{V_A}{V_{bi}} \right]^m}, \qquad 1/3 \le m \le 1/2; \; V_A \le 0 \tag{5.8}$$

For a linearly graded junction $m = 1/3$; for a step junction $m = 1/2$. Most real junctions, fabricated by standard procedures, are found to have an m of between $1/3$ and $1/2$.

It should be noted that Eq. (5.8) is applicable to small forward-biased voltages of less than about $V_{bi}/2$. Figure 5.3 illustrates that when the diode is forward biased, C_J increases very rapidly and then decreases [not part of Eq. (5.8)].

The applications of junction capacitors are extensive. In bipolar integrated circuits the reverse-biased p-n junction is used to isolate the transistors and resistors from each other. Almost every FM tuner and TV set uses the dc voltage-variable capacitor in automatic tuning circuits; it has no moving parts and the capacitance changes in response to a dc level.

5.1.2   Conductance

The small-signal conductance of the reverse-biased junction is derived under the assumption that the carriers are able to respond to the signal quasi-statically; that is, the carriers return to near steady state in much less time than the period of the signal. In the case of reverse bias this means that signal frequencies are less than about 100 MHz for most devices, since the carriers responding are the majority carriers.

Under the assumption of majority carrier quasi-static response, the diode reacts instantaneously to a signal superimposed on the dc operating variables. The diode can be represented by the ideal diode equation; that is, if the diode current is a function of the dc voltage V_A that is perturbed by a small signal v_a, Eq. (5.9) is modified as per Eq. (5.10).

$$I(V_A) = I_0[e^{qV_A/kT} - 1] \tag{5.9}$$

$$I(V_A + v_a) = I_0[e^{q(V_A + v_a)/kT} - 1] \tag{5.10}$$

and the signal current, i, is defined as Eq. (5.11),

$$i = I(V_A + v_a) - I(V_A) \tag{5.11}$$

Equation (5.10) can be expanded into the Taylor series of Eq. (5.12) and, because $|v_a| \ll V_A$, only the first two terms need be retained, giving Eq. (5.13),

$$f(x + h) = f(h) + x\frac{df}{dh} + \cdots \tag{5.12}$$

$$I(v_a + V_A) \cong I(V_A) + v_a\frac{dI}{dV_A} \tag{5.13}$$

Therefore the signal current is obtained from Eq. (5.11) and Eq. (5.13), as

$$i = v_a \frac{dI}{dV_A} = \left[\frac{dI}{dV_A} \right] v_a \tag{5.14}$$

The low-frequency ($\omega \to 0$) conductance is defined from Eq. (5.14) as

$$G_0 = \frac{i}{v_a} = \frac{dI}{dV_A} \quad \text{(mhos)} \tag{5.15}$$

By differentiating the ideal diode equation, Eq. (5.9), we obtain the low-frequency conductance as Eq. (5.16),

$$G_0 = \frac{dI}{dV_A} = I_0 \frac{q}{kT} e^{qV_A/kT} = \frac{q}{kT} (I + I_0) \tag{5.16}$$

The last form of Eq. (5.16) was obtained by adding I_0 to both sides of Eq. (5.9). A common model element is the *dynamic resistance*, defined as the reciprocal of G_0,

$$\boxed{ r = \frac{1}{G_0} = \frac{kT}{q(I + I_0)} } \quad \text{(ohm)} \tag{5.17}$$

Note that for several tenths of a volt of reverse bias, $I \to -I_0$ and $r \to \infty$.

For the reverse bias case of a silicon diode at room temperature, as discussed in Chapter 4, the generation current in W dominates conduction. Applying Eq. (5.15) to that case yields

$$G_0 = \frac{q}{kT} (I + I_0) - \frac{qA}{2\tau_0} n_i \frac{dW}{dV_A} \tag{5.18}$$

Differentiating Eq. (2.51) results in Eq. (5.19),

$$\frac{dW}{dV_A} = \left[\frac{2K_S \varepsilon_0}{q} \frac{(N_A + N_D)}{N_A N_D} \right]^{1/2} \frac{(1/2)(-1)}{(V_{bi} - V_A)^{1/2}} \tag{5.19}$$

The low-frequency reverse bias conductance is then given by Eq. (5.20),

$$G_0 = \frac{q}{kT} (I + I_0) + \frac{qAn_i}{4\tau_0} \left[\frac{2K_S \varepsilon_0}{q} \frac{(N_A + N_D)}{N_A N_D} \frac{I}{(V_{bi} - V_A)} \right]^{1/2} \tag{5.20}$$

Note that the nonideal silicon diode has a finite conductance that depends on the generation current originating in the depletion region. Equation (5.16) and Eq. (5.20) for G_0 are also interpreted as the slope of the $I - V_A$ curve. Any phenomenon that increases the slope increases the conductance. One final observation is that Eq. (5.16) will be shown in the next section to be valid for forward bias. The validity is, however, at frequencies considerably lower than for the reverse bias case.

SEE EXERCISE 5.1 – APPENDIX A

5.2   FORWARD-BIASED JUNCTION ADMITTANCE

The junction diode under forward bias conditions ($V_A > 0$) and perturbed by a small-signal sinusoid is also modeled by an admittance. In addition to the depletion region capacitance C_J (which is a result of majority carrier response), the minority carrier response yields a *diffusion capacitance* C_D. The minority carriers also contribute to the conductance (G) of the junction admittance. Figure 5.5 illustrates the signal equivalent circuit where R_s is the series resistance of the diode due to the bulk regions and/or the ohmic contacts. In general, the conductance and diffusion capacitance are a function of the signal frequency and the dc operating point variables, as will become evident by the following derivation. Let's consider the response of the minority carriers.

The minority carrier concentrations for *a p-n* junction forward biased with a dc voltage $V_A > 0$ and perturbed by a small-signal sinusoid are illustrated in Fig. 5.6. The average minority carrier concentration is a result of the voltage V_A across the depletion region. Increments above and below the average are due to the signal v_a applied to the depletion region edges. As v_a becomes positive and negative the carrier distributions at x_n and $-x_p$ change. Because diffusion is a relatively slow process compared with the signal frequency, the charge distribution increment propagates into the bulk region, fluctuating

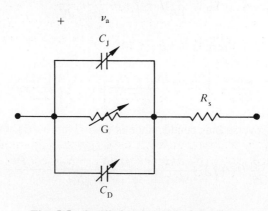

Fig. 5.5   Small-signal model of the diode.

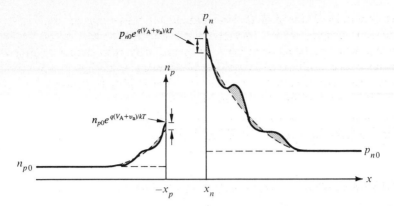

Fig. 5.6 High-frequency charge-storage p^+-n diode at one instant of time.

above and below the average (dc) carrier distribution. Figure 5.6 illustrates the fluctuation of the electrons and holes for one instant of time.

A dynamic picture of Fig. 5.6 can be conceptualized by visualizing that you are holding one end of a long rope. When held fixed in position it drops from your hand to the floor in a manner similar to the dc steady-state, minority carrier distribution. If the held end of the rope is oscillated rapidly, the rope propagates the pulsation down its length until the pulse eventually decays away on the floor.

The diode minority carrier distribution propagates a signal current in response to the signal voltage. We should therefore expect an associated junction admittance that has a conductance and a susceptance, both of which will be functions of the dc operating point and the signal frequency. Figure 5.6 illustrates that the minority carrier distributions are a function of time and space; that is, $p_n(x, t)$.

To derive the diffusion admittance of a junction (the minority carrier response), let us assume a p^+-n junction and therefore concern ourselves only with $p_n(x, t)$. A similar derivation holds for $n_p(x, t)$. The minority carrier continuity equation for a uniformly doped n-region is the starting point, repeated here for the reader's convenience from Vol. I, Chapter 3:

$$\frac{\partial \Delta p_n(x, t)}{\partial t} = D_P \frac{\partial^2 \Delta p_n(x, t)}{\partial x^2} - \frac{\Delta p_n(x, t)}{\tau_p} \qquad (5.21)$$

Figure 5.6 indicates that the signal oscillates the hole distribution around its average (dc) value. This allows the solution to be broken up into a solution for the dc component plus the signal components, as indicated by Eq. (5.22):

$$\Delta p_n(x, t) = \overline{\Delta p_n}(x) + \tilde{p}_N(x, t) \qquad (5.22)$$

Substituting Eq. (5.22) into Eq. (5.21) yields Eq. (5.23),

$$\frac{\partial \overline{\Delta p}_n(x)}{\partial t} + \frac{\partial \tilde{p}_N(x, t)}{\partial t} = D_P \frac{\partial^2 \overline{\Delta p}_n(x)}{\partial x^2} + D_P \frac{\partial^2 \tilde{p}_N(x, t)}{\partial x^2} - \frac{\overline{\Delta p}_n(x)}{\tau_p} - \frac{\tilde{p}_N(x, t)}{\tau_p}$$

$$(5.23)$$

Since $\overline{\Delta p}_n(x)$ is not a function of time, the first term of Eq. (5.53) is equal to zero. Separately equating the coefficients of the average terms and the signal terms results in Eqs. (5.24) and (5.25).

$$0 = D_P \frac{\partial^2 \overline{\Delta p}_n(x)}{\partial x^2} - \frac{\overline{\Delta p}_n(x)}{\tau_p} \qquad (5.24)$$

$$\frac{\partial \tilde{p}_N(x, t)}{\partial t} = D_P \frac{\partial^2 \tilde{p}_N(x, t)}{\partial x^2} - \frac{\tilde{p}_N(x, t)}{\tau_p} \qquad (5.25)$$

Equation (5.24) was solved in Chapter 3 and results in the dc solution, in that chapter, led to the I versus V_A relationship known as the ideal diode equation. Let us turn our attention to the solution of Eq. (5.25), which will lead us to the small-signal admittance, our ultimate goal.

An examination of Eq. (5.25) shows it to be a type of differential equation that can be solved as a product solution (similar to the wave equation in electromagnetic field theory). The solution is broken into the signal part that is only a function of x and the signal part that is only a function of time.

$$\tilde{p}_N(x, t) = \hat{p}_N(x) f(t) \qquad (5.26)$$

If we assume the forcing function to be a sine or cosine function, then $f(t) = e^{j\omega t}$, similar to the phasor concept from circuit theory,

$$\tilde{p}_N(x, t) = \hat{p}_N(x) e^{j\omega t} \qquad (5.27)$$

Therefore, if Eq. (5.27) is an assumed solution it must satisfy Eq. (5.25); that is, performing the differentiations of Eq. (5.25), using Eq. (5.27), must yield

$$\frac{\partial \tilde{p}_N(x, t)}{\partial t} = j\omega \hat{p}_N(x) e^{j\omega t} \qquad (5.28)$$

$$\frac{\partial^2 \tilde{p}_N(x, t)}{\partial x^2} = \frac{d^2 \hat{p}_N(x)}{dx^2} e^{j\omega t} \qquad (5.29)$$

Substitution into Eq. (5.25) gives Eq. (5.30), and cancelling the $e^{j\omega t}$ terms before collecting the $\hat{p}_n(x)$ terms yields Eq. (5.31),

$$j\omega\hat{p}_N(x)e^{j\omega t} = D_P\frac{d^2\hat{p}_N(x)}{dx^2}e^{j\omega t} - \frac{\hat{p}_N(x)e^{j\omega t}}{\tau_p} \tag{5.30}$$

or

$$j\omega\hat{p}_N(x) + \frac{\hat{p}_N(x)}{\tau_p} = D_P\frac{d^2\hat{p}_N(x)}{dx^2} = \hat{p}_N(x)\left[\frac{1}{\tau_p} + j\omega\right] \tag{5.31}$$

Dividing by D_P results in an equation very similar to the differential equation for the dc solution of the minority carrier concentrations,

$$\frac{d^2\hat{p}_N(x)}{dx^2} = \hat{p}_N(x)\left[\frac{1 + j\omega\tau_p}{D_P\tau_p}\right] \tag{5.32}$$

To make Eq. (5.32) have the same form as the dc case, we define the complex diffusion length for holes as Eq. (5.33),

$$[L_P^*]^2 = \frac{D_P\tau_p}{1 + j\omega\tau_p} = \frac{L_P^2}{1 + j\omega\tau_p} \tag{5.33}$$

Then Eq. (5.32) can be written as Eq. (5.34),

$$\frac{d^2\hat{p}_N(x)}{dx^2} = \frac{\hat{p}_N(x)}{[L_P^*]^2} \tag{5.34}$$

The solution of Eq. (5.34), by analogy with the dc solution of Eq. (5.24) [or Eq. (3.36)], is

$$\hat{p}_N(x) = B_1 e^{x/L_P^*} + B_2 e^{-x/L_P^*} \tag{5.35}$$

As a boundary condition we know that $|\hat{p}_N(x)|$ cannot grow without bound; that is, as $x \to \infty$ $|\hat{p}_N(x)| \to 0$, and therefore B_1 must be zero. At the depletion region edge, the signal component of the hole distribution is controlled by the applied signal voltage. To evaluate the constant B_2 we need to know the value of $\hat{p}_N(0)$. The total carrier distribution is determined by the total voltage applied as given by Eq. (5.36), which is the same as Eq. (3.30), with $V_A \to V_A + v_a(t)$ and the x' axis starting at x_n,

$$p_n(0, t) = p_{n0}\exp\left[\frac{q(V_A + v_a(t))}{kT}\right] \tag{5.36}$$

The signal is small; that is, $|v_a(t)| \ll V_A$, and the series expansion of the exponential can be applied to Eq. (5.36) as

$$e^x \cong 1 + x + \cdots, \qquad \text{for } x \ll 1 \tag{5.37}$$

or

$$p_n(0, t) \cong p_{n0}e^{qV_A/kT}\left[1 + \frac{qv_a(t)}{kT}\right]$$

$$= p_{n0}e^{qV_A/kT} + p_{n0}e^{qV_A/kT}\frac{qv_a(t)}{kT} \tag{5.38}$$

Note that the signal part is the last term of Eq. (5.38), and therefore

$$\hat{p}_n(0) = p_{n0}e^{qV_A/kT}\frac{qv_a}{kT} = B_2 \tag{5.39}$$

where the constant B_2 was evaluated using Eq. (5.35).

The signal current is obtained from the diffusion current formula of Eq. (5.40),

$$i = -qAD_P\frac{d\hat{p}_N(x)}{dx}\bigg|_{x=0} = \frac{qAD_P}{L_P^*}p_{n0}e^{qV_A/kT}\frac{qv_a}{kT} \tag{5.40}$$

The junction admittance is then

$$Y_p = A\frac{q}{kT}\left[q\frac{D_P}{L_P^*}p_{n0}\right]e^{qV_A/kT} \tag{5.41}$$

From Eq. (5.33)

$$L_P^* = \frac{L_P}{\sqrt{1 + j\omega\tau_p}} \tag{5.42}$$

and Eq. (5.41) can be rewritten as Eq. (5.43),

$$Y_p = \frac{qA}{kT}\left[q\frac{D_P p_{n0}}{L_P}\sqrt{1 + j\omega\tau_p}\right]e^{qV_A/kT} \tag{5.43}$$

Note that the square root of a complex number is another complex number, and Eq. (5.43) will have a real part which is the conductance G and an imaginary part which will be the capacitive susceptance ωC_D. The *diffusion capacitance* is then defined as

$$C_D = \frac{\text{Imaginary part of } Y}{\omega} \tag{5.44}$$

Equation (5.43) can easily be extended to include the minority carrier electron contribution by using the idea of "complements" introduced earlier. The full derivation for electrons is identical to that for the holes. Therefore the total diffusion admittance of a *p-n* junction is

$$Y = \frac{qA}{kT}\left[q\frac{D_P}{L_P}p_{n0}\sqrt{1 + j\omega\tau_p} + q\frac{D_N}{L_N}n_{p0}\sqrt{1 + j\omega\tau_n}\right]e^{qV_A/kT} \qquad (5.45)$$

where the real part of Eq. (5.45) is the junction conductance due to holes in the *n*-region and electrons in the *p*-region. Similarly, the diffusion capacitance, due to holes and electrons, is obtained from the imaginary part of Eq. (5.45) divided by ω.

SEE EXERCISE 5.2 – APPENDIX A

5.3 LIMITING CASES

5.3.1 *p*$^+$-*n* Diode Admittance

To develop a number of important concepts and practical results, let us return to a simpler case, that of the p^+-*n* device. The doping asymmetry of the heavily doped region makes $n_{p0} \ll p_{n0}$, and the second term inside the bracket of Eq. (5.45) becomes insignificant. If $\omega \rightarrow 0$, Eq. (5.43) becomes the definition of the low-frequency conductance, G_0,

$$Y\big|_{\omega\rightarrow0} = \frac{q}{kT}\left[qA\frac{D_P p_{n0}}{L_P}e^{qV_A/kT}\right] = G_0 \qquad (5.46)$$

Now Eq. (5.43) can be simplified to Eq. (5.47),

$$Y \cong G_0\sqrt{1 + j\omega\tau_p} \qquad (5.47)$$

To evaluate Eq. (5.47), it is necessary to take the square root of a complex number. This means converting $1 + j\omega\tau_p$ to polar coordinates, taking the square root of the magnitude and one-half the angle, followed by a conversion back to rectangular coordinates, yielding a real part and an imaginary part.

The frequency dependence of G and C_0 is most easily observed by plotting Eq. (5.47) for specific values of $\omega\tau_p$. Figure 5.7 illustrates that the conductance begins to increase

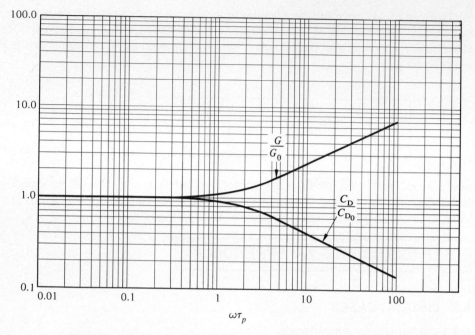

Fig. 5.7 Frequency dependence of G and C_D for a p^+-n junction.

with frequency, above the low-frequency value, when $\omega\tau$ is about 0.5, while the diffusion capacitance decreases at the higher frequencies.

The dependence of the conductance and diffusion capacitance on the dc operating point variable is most easily obtained from the ideal diode equation and Eq. (5.47). For any reasonable degree of forward bias

$$G \propto e^{qV_A/kT} \propto I \tag{5.48}$$

and

$$C_D \propto G_0 \propto I \tag{5.49}$$

The low-frequency conductance G_0 of Eq. (5.46) is obtained as follows. Note that the bracketed term of Eq. (5.46) contains I_0 from the ideal diode equation for a p^+-n device,

$$\frac{qAD_P p_{n0}}{L_P} e^{qV_A/kT} = I_0 e^{qV_A/kT} = (I + I_0) \tag{5.50}$$

therefore

$$G_0 = \frac{q}{kT}(I + I_0) = \frac{dI}{dV_A} \tag{5.51}$$

From Fig. 5.7, this means that Eq. (5.5) is applicable for $\omega\tau_p < 0.5$ for both forward and reverse bias.

5.3.2 p^+-n Diode, $\omega\tau_p \ll 1$

Equation (5.47) can be expanded in a Taylor series of Eq. (5.12) if $\omega\tau_p \ll 1$, the result of which is Eq. (5.52),

$$Y \cong G_0\left[1 + \frac{j\omega\tau_p}{2}\right] = G_0 + j\omega\frac{G_0\tau_p}{2}, \qquad \text{for } \omega\tau_p \ll 1 \tag{5.52}$$

and the diffusion capacitance at low frequencies becomes

$$\boxed{C_{D_0} \cong \frac{G_0\tau_p}{2}} \tag{5.53a}$$

$$C_{D_0} = \frac{qI}{kT}\left[\frac{\tau_p}{2}\right] \tag{5.53b}$$

which is frequency independent as illustrated by Fig. 5.7. Again note the dependence of C_D on $G_0 \propto I$. Also, the larger τ_p, the more minority carrier charge-storage phenomena; that is, a larger C_D.

5.3.3 Series Resistance

Most practical devices have some small but finite resistance in the bulk regions. This resistance is in series with the device. The ohmic contacts (the metal to silicon) also add resistance. Figure 5.5 illustrates the sum of these two as R_s. It should be noted that R_s is fundamentally the same resistance as discussed in Chapter 4, at large current levels, where the real device deviated from the ideal.

5.3.4 *p-n$^+$* Diode

The idea of complements indicates that the *p-n$^+$* junction admittance is

$$Y_n \cong G_0\sqrt{1 + j\omega\tau_n}$$

(5.54)

where Fig. 5.7 can be used by replacing $\omega\tau_p$ with $\omega\tau_n$. Similarly,

$$G_0 = \frac{q}{kT}\left[qA\frac{D_N n_{p0}}{L_N}\right]e^{qV_A/kT} \cong \frac{qI}{kT}\left[\frac{\tau_n}{2}\right]$$

(5.55)

and as $\omega\tau_n \ll 1$, then the diffusion capacitance is approximated as

$$C_{D_0} \cong \frac{G_0\tau_n}{2}, \quad \text{for } \omega\tau_n \ll 1$$

(5.56)

which is independent of frequency but proportional to the dc current I.

5.4 SUMMARY

The junction admittance for the reverse-biased diode was developed by considering the superimposed, small signal (v_a) on the dc voltage and its effect on the current. At frequencies where the majority carriers have ample time to respond to the signal voltage, the junction was modeled by a *junction depletion capacitance* and a conductance. The depletion capacitance decreased with larger reverse biases, as did the conductance. An increase in doping densities, N_A and/or N_D, increased the junction capacitance by making the average depletion width smaller.

The case of the diode in forward bias resulted in a diffusion admittance that is dependent on the signal frequency. At very low frequencies the conductance and diffusion capacitance are constant. For frequencies such that $\omega\tau$ approaches 0.5, the conductance increases, and the diffusion capacitance decreases with increasing frequency. The admittance increases exponentially with V_A and is therefore proportional to the dc current I. Both G and C_D are therefore proportional to the dc operating point current.

The total forward-biased small-signal equivalent circuit contains C_J, G, and C_D in parallel, with R_s in series with the combination.

PROBLEMS

5.1 A step junction is doped $N_A = 10^{+17}/cm^3$ and $N_D = 10^{+15}/cm^3$. Let $kT = 0.026$ eV, $A = 10^{-5}$ cm^2 and $n_i = 10^{+10}/cm^3$.

(a) Calculate the depletion capacitance at $V_A = 0$, C_{J0}.

(b) Calculate the depletion capacitance C_J at $V_A = -1$ and -10 volts.

(c) Calculate C_{J0} if $N_D = 10^{+16}/cm^3$.

(d) What is the ratio of part (c) to part (a)? Is it a square-root dependence on N_D?

5.2 (a) Using the results of Chapter 2, derive an equation for the depletion capacitance C_J of a linearly graded junction.

(b) If $a = 5 \times 10^{+19}/cm^4$, $n_i = 10^{+10}/cm^3$, $kT = 0.026$ eV, and $A = 10^{-4}$ cm^2, calculate the depletion capacitance C_J if

   (i) $V_A = 0$,

   (ii) $V_A = -2$,

   (iii) $a = 10^{20}/cm^4$, and $V_A = 0$.

(c) What can be generalized from the results of part (b)?

5.3 The junction depletion capacitance for an abrupt junction has the measured values of Table P5.3.

(a) Carefully plot $1/C^2$ vs. V_A and determine C_{J0} and V_{bi} from the graph.

(b) Outline the procedure for a linearly graded junction.

Table P5.3

C (pF)	V_A
3.993	−0.5
3.420	−1.0
2.764	−2.0
2.381	−3.0
2.123	−4.0
1.934	−5.0

5.4 A p-n abrupt junction with $N_A = 10^{+17}/cm^3$ and $N_D = 5 \times 10^{+15}/cm^3$ has $\tau_p = 0.1$ μs and $\tau_n = 0.01$ μs with $kT = 0.026$ eV, $n_i = 10^{+10}/cm^3$, $A = 10^{-4}$ cm^2, $\mu_n = 801$ cm^2/V-s, and $\mu_p = 438$ cm^2/V-s.

(a) Calculate the junction depletion capacitance at

   (i) $V_A = 0$,

   (ii) $V_A = -V_{bi}/2$,

   (iii) $V_A = -10$ volts

(b) Calculate the junction diffusion capacitance ($\omega\tau < 0.1$ for both the n- and the p-region) at

   (i) $V_A = V_{bi}/2$,

   (ii) $V_A = 0.9\ V_{bi}$.

   (iii) Calculate the junction conductance at $V_A = V_{bi}/2$ and $V_A = 0.9\ V_{bi}$.

(c) Discuss which carrier type and region dominates the diffusion capacitance and the junction capacitance. Is the diffusion capacitance proportional to the forward-biased current?

5.5 If $\omega = 10^7$ rad/s in Problem 5.4, calculate the diffusion capacitance, C_D, and the conductance at $V_A = 0.9\ V_{bi}$.

5.6 For the p^+-n step junction sketch plots of:

(a) C_{J0} vs. N_D;

(b) C_D vs. τ_p when $\omega\tau_p < 0.1$;

(c) C_D vs. τ_p when $\omega\tau_p = 10$ and τ_p is increased by a factor of 10. By what ratio did the diffusion capacitance change? The conductance?

5.7 A p^+-n step junction has $\tau_p = 1.0\ \mu$s. If $\omega = 10^6$ rad/s, sketch a plot and indicate the slopes of

(a) G vs. I;

(b) C_D vs. I;

(c) if $\omega = 10^5$ rad/s repeat parts (a) and (b).

(d) Do the results of parts (a), (b), and (c) agree with Fig. 5.7?

6 / Switching Response

The *p-n* junction has many applications in which the device is used as an electrical switch. Typically, a pulse of current or voltage is used to switch the diode from reverse bias, called the "off" state, to forward bias, called the "on" state, and vice versa. Of prime concern to the circuit and device designer are the speeds at which the *p-n* junction can be made to switch states. We will first qualitatively discuss the transient which occurs when the diode is switched from the on to the off state and explain the origins of the time delay before the device turns off. The *reverse recovery time* is defined as the time delay from initiation of switching until the diode recovers to a specified degree of off, that is, a level of reverse current. The largest component of the reverse recovery time is the *storage time*. An approximate solution for the storage time is derived and compared to a more detailed mathematical solution. The transient response from reverse bias to forward bias is also discussed and an approximate solution is derived for one specific "turn-on" case. It should be noted that many of the concepts developed in this chapter apply to the bipolar transistor of Volume III.

6.1 THE TURN-OFF TRANSIENT

The dynamic switching of a diode from the forward conduction, or on, state to the off state is known as the turn-off transient. For the diode to be an *ideal* switch in the turn-off cycle, the current would traverse instantly from I_F, the forward dc value, to the reverse leakage current $-I_0$. It should be apparent from the discussions of Chapter 3, however, that the minority carrier charge stored in the bulk *n*- and *p*-regions must be removed before the device can be switched from forward conducting to reverse bias. This requires instantaneous removal of charge, the minority carriers in the bulk regions. Therefore, the real device cannot function as an ideal switch.

Figure 6.1(a) illustrates the idealized experiment for measuring the turn-off time of the diode and Fig. 6.1(b) is a sketch of the current response. The reverse current $-I_R$ is defined as the current an instant after the device is switched, while the *reverse recovery time* (t_{rr}) is defined as the time necessary for the diode current to recover to $-0.1\,I_R$. Figure 6.1(c) defines the *storage time* (t_s) as the time when the junction voltage $v_A(t)$ reaches zero volts or, as shown in Fig. 6.1(b), the time associated with the nearly con-

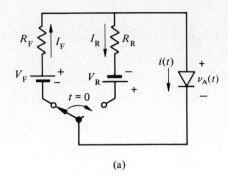

(a)

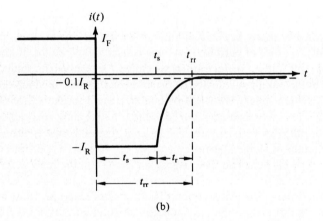

(b)

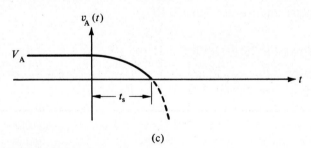

(c)

Fig. 6.1 Reverse recovery time and storage time: (a) circuit; V_F and V_R are much larger than $|v_A(t)|$; (b) current; (c) voltage.

stant current $(-I_R)$ of Fig. 6.1(b). The storage time, as will be discussed later, increases with the amount of stored minority charge and is used as a figure of merit in switching applications. The *recovery time* (t_r) is the difference between t_{rr} and t_s.

The dc forward current (I_F) is obtained by inspection of Fig. 6.1(a) at $t = 0^-$, where V_F is assumed to be much greater than the dc diode voltage V_A. Since V_A is typically between 0.1 and 0.875 volts, the criterion is easily met in most cases where $V_F \gtrsim 20$ volts. Writing a loop equation around the circuit, solving for the dc current, and assuming $V_A \ll V_F$ yields Eq. (6.1):

$$I_F = \frac{V_F - V_A}{R_F} \cong \frac{V_F}{R_F} \qquad (6.1)$$

The reverse current (I_R) has a similar definition for $t = 0^+$, where the condition of $|v_A(t)| \ll V_R$ must be satisfied.

$$I_R = \frac{V_R + v_A}{R_R} \cong \frac{V_R}{R_R} \qquad (6.2)$$

It should be noted that it is possible to make I_R much larger than I_F, a technique often used by switching-circuit designers to reduce the switching times.

The reader may initially be somewhat puzzled by Fig. 6.1(b), in which a large reverse current momentarily flows through the diode. Let's consider the physical explanation of the phenomenon before treating it analytically. The minority carrier concentrations before the device is switched are shown in Fig. 6.2(a). The total excess minority carrier charge stored in the junction is the area under each curve above p_{n0} and n_{p0}, respectively. Long after switching, at $t \to t_{rr}$, the minority carrier distributions are less than their thermal equilibrium values, as shown in Fig. 6.2(b). Obviously the charge transfer necessary

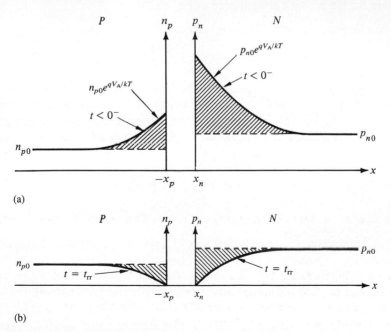

Fig. 6.2 (a) Minority carrier charge storage; (b) majority carrier depletion.

to turn the device off is represented by the sum of the cross-hatched areas. Remember that the minority carrier concentrations change in the quasi-neutral bulk regions by diffusing (leaving) and/or recombination. Therefore, to change the minority carrier concentrations of Fig. 6.2 from their $t = 0^-$ to $t \to t_{rr}$ values, current flow and recombination are necessary. Current flow, $-I_R$, is a result of minority carrier holes leaving the n-region and minority carrier electrons leaving the p-region by diffusion. Recombination occurs wherever the carrier concentrations exceed their thermal equilibrium values.

It should be apparent that the larger the minority carrier injected charge to be removed from the bulk regions, the longer the turn-off transient.

Figure 6.2 can be used to deduce several qualitative relationships between the turn-off transient and the material properties. First, if τ_p and τ_n are decreased (L_N and L_P are reduced), then less minority charge is stored at $t = 0^-$, and the turn-off time is made shorter because less charge transfer is necessary. Secondly, shorter lifetimes mean larger recombination of excess carriers, and hence a shortened turn-off time due to quicker charge removal. The circuit variables for shorter turn-off times are also closely coupled with the stored charge. A larger I_F produces a larger stored charge at $t = 0^-$ because a larger V_A is required and there is therefore a larger minority carrier concentration. Similarly, a larger I_R produces a quicker removal of charge and hence a shorter turn-off time.

Now let's consider how the minority carriers change with time and distance as the device is switched from on to off. The derivation for dc currents in an ideal diode assumed that no generation or recombination occurred in the depletion W. With this assumption the terminal current was obtained by adding Eq. (3.14) to Eq. (3.15), repeated here for the reader's convenience as Eq. (6.3).

$$ J = -qD_P \frac{dp_n}{dx}\bigg|_{x_n} + qD_N \frac{dn_p}{dx}\bigg|_{-x_p} \tag{6.3} $$

Equation (6.3) is interpreted as stating that the slope of the minority carrier concentration, evaluated at the depletion region edge, determines the value of each current component. Also, remember that the minority carrier concentrations at the depletion region edges were determined from Eqs. (3.30) and (3.28), repeated as Eqs. (6.4) and (6.5) with V_A replaced by $v_A(t)$.

$$ p_n(x_n) = p_{n0} e^{qv_A(t)/kT} = p_n(x_n, t) \tag{6.4} $$

$$ n_p(-x_p) = n_{p0} e^{qv_A(t)/kT} = n_p(-x_p, t) \tag{6.5} $$

With the above equations in mind we are now in a position to explain the general nature of the switching response of the experiment shown in Fig. 6.1.

Figure 6.3 illustrates the time progression of excess charge removal with I_R at a constant value, that is, with a constant slope of the carrier concentrations at the depletion region edges: positive slope for holes (negative current) and negative slope for electrons (negative current). The value of I_R is nearly a constant as long as $|v_A(t)| \ll V_R$.

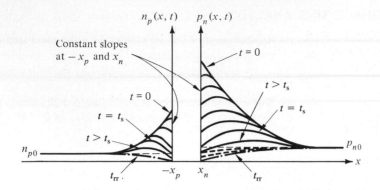

Fig. 6.3   Turn-off transient carrier concentrations.

For the excess holes in the n-region, the constant diffusion current at x_n removes enough carriers to cause $p_n(x_n, t)$ to decrease in magnitude and there is hence a reduction in the terminal voltage according to Eq. (6.4). This reduction of $v_A(t)$ is illustrated in Fig. 6.1(c). Similar arguments for $n_p(-x_p, t)$ reiterate the reduction in excess carriers and the terminal voltage.

The storage time (t_s) is defined as occurring at the end of the constant current phase of the reverse current transient. Physically, the constant current phase ends when $|v_A(t)|$ becomes comparable to V_R and the two voltages subtract from each other, forcing $i(t)$ to be less negative than $-I_R$, as is illustrated in Fig. 6.1(b). Figure 6.3 shows t_s as the time at which the junction voltage has become zero; that is, $p_n(x_n) = p_{n0}$ at $t = t_s$.

The remainder of the reverse current transient (t_r) is characterized by $v_A(t)$ becoming large and negative; that is, a reverse voltage develops across the junction. The current $i(t)$ is no longer a constant and is decreasing rapidly towards the steady-state, reverse saturation current $-I_0$. During this phase of the recovery the excess charge is being removed primarily by recombination (although some diffusion persists). Ultimately, the total excess charge is eliminated and the reverse bias causes the carrier deficit shown in Fig. 6.3 at $t = t_{rr}$. By the end of the reverse transient, $v_A \cong -V_R$ and the device approaches the steady-state dc reverse bias condition.

The previous discussions allow us to predict some general results. The larger the forward current I_F the larger the minority carrier charge storage. Therefore, with a fixed I_R this means a larger t_s. A larger value of τ_p and τ_n means larger L_P and L_N values, resulting in more stored charge and again a larger t_s. In some fast-switching silicon diodes, the device is "gold doped" in addition to the n- and p-type impurities. Gold acts as a very effective recombination center near E_i in the band gap of silicon and thereby reduces the minority carrier lifetimes τ_n and τ_p, which in turn reduces t_s. Generalizing then, the larger the excess minority charge storage, the longer it takes to discharge the junction. Finally, the storage time is also reduced by increasing the value of the reverse current I_R, pulling more charge per second from the bulk regions.

6.2 STORAGE TIME ANALYSIS

An approximate solution for the storage time can be obtained for the ideal step junction diode, confirming many of our previous predictions. For simplicity, assume a p^+-n junction where the stored minority carrier charge is dominated by the holes in the bulk n-region; i.e., $Q_P \gg Q_N$. For this type of device, as discussed in Chapter 3, the total dc current is approximately the hole current evaluated at x_n. Figure 6.4 illustrates the hole concentration, initially, and for the phases of the reverse recovery of the p^+-n diode.

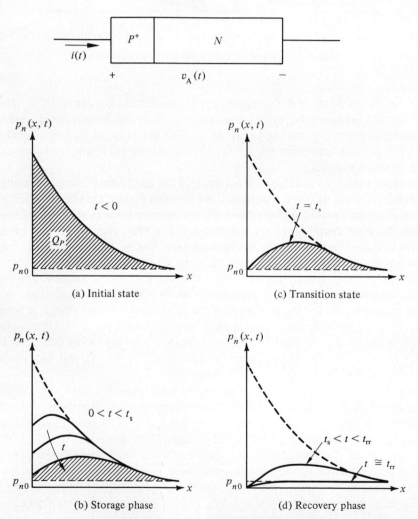

Fig. 6.4 Reverse recovery hole concentration profiles: (a) initial; (b) storage phase; (c) $t = t_s$; (d) recovery phase. Remember that these are "log" plots on the vertical axis.

The fundamental problem requires a solution for $\Delta p_n(x, t)$ as sketched in Fig. 6.4, a difficult problem indeed. We sidestep the problem by considering the total injected minority carrier charge $Q_P(t)$, remembering that the charge can change only through current flow $i(t)$ and recombination.

Volume I, Chapter 3, considered the *minority* carrier continuity equations for the bulk regions, repeated here for the n-bulk region, assuming low-level injection.

$$\frac{\partial \Delta p_n}{\partial t} = -\frac{1}{q}\frac{\partial J_P}{\partial x} - \frac{\Delta p_n}{\tau_p} \tag{6.6}$$

Equation (6.6) is multiplied by the area (A) and by q, then integrated over the n-bulk region.

$$\frac{d}{dt}\left[qA \int_{x_n}^{\infty} \Delta p_n(x, t)\, dx \right] = -A \int_{J_P(x_n)}^{J_P(\infty)} dJ_P - \frac{1}{\tau_p}\left[(qA) \int_{x_n}^{\infty} \Delta p_n(x, t)\, dx \right] \tag{6.7}$$

It must be noted that the total hole charge is Eq. (6.8).

$$Q_P(t) = qA \int_{x_n}^{\infty} \Delta p_n(x, t)\, dx \tag{6.8}$$

Equation (6.7) can be rewritten as Eq. (6.9), after recognizing the $Q_P(t)$ terms of Eq. (6.8) as they appear in Eq. (6.7).

$$\frac{d}{dt}[Q_P(t)] = -A[J_P(\infty) - J_P(x_n)] - \frac{Q_P(t)}{\tau_p} \tag{6.9}$$

Remember that for the "long-base diode" the n-region is infinite in length; therefore $J_P(\infty)$ is zero since $d\Delta p_n/dx\,|_\infty = 0$. For the p^+-n junction the total current $i(t)$ is approximately $J_P(x_n)$; hence Eq. (6.9) becomes Eq. (6.10).

$$\boxed{\frac{dQ_P(t)}{dt} = i(t) - \frac{Q_P(t)}{\tau_p}} \tag{6.10}$$

Examination of Eq. (6.10) reveals that the rate of change of charge storage is equal to the current (adding or subtracting charges per second) minus that lost to recombination. The total stored-hole charge changes by addition or removal of holes due to current and by recombination.

Let's now apply Eq. (6.10) to the approximate solution for the storage time of a p^+-n junction, using the circuit constraints of Fig. 6.1(a). For $t \geq 0$, but less than t_s, the current $i(t) = -I_R$ in Eq. (6.10), as written in Eq. (6.11).

$$\frac{dQ_p(t)}{dt} = -I_R - \frac{Q_p(t)}{\tau_p}, \qquad 0^+ \leq t \leq t_s \qquad (6.11)$$

Equation (6.11) can be solved directly as a differential equation or by applying a Laplace transform. It is a variables-separable type of differential equation. Separating the variables by multiplying by dt, then dividing by $I_R + Q_p(t)/\tau_p$ and integrating, we obtain Eq. (6.12a), where $Q_p(t_s)$ is assumed to be approximately zero,

$$\int_{Q_p(0^+)}^0 \frac{dQ_p}{\left[I_R + \dfrac{Q_p(t)}{\tau_p}\right]} = -\int_0^{t_s} dt = -t \Big|_0^{t_s} = -t_s \qquad (6.12a)$$

Figure 6.4(c) shows that $Q_p(t_s) \neq 0$; however, to assume that it is zero results in a conservative estimate of t_s (a larger value). The term $Q_p(0^+)$ is illustrated in Fig. 6.4(a) as the initial hole charge stored in the bulk n-region. The left side of Eq. (6.12a) can be integrated from a standard table of integrals to be

$$\tau_p \ln\left(I_R + \frac{Q_p(t)}{\tau_p}\right)\Big|_{Q_p(0^+)}^0 = -t_s = \tau_p \ln\left[I_R + \frac{0}{\tau_p}\right] - \tau_p \ln\left[I_R + \frac{Q_p(0^+)}{\tau_p}\right]$$

$$(6.12b)$$

Solving for t_s yields Eq. (6.12c),

$$t_s = -\tau_p \ln I_R + \tau_p \ln\left[I_R + \frac{Q_p(0^+)}{\tau_p}\right] \qquad (6.12c)$$

therefore

$$t_s = \tau_p \ln\left[1 + \frac{Q_p(0^+)}{\tau_p I_R}\right] \qquad (6.13)$$

Although $Q_p(t_s) \neq 0$ and, compared to $Q_p(0^+)$, it may not be entirely negligible, the approximation nevertheless yields reasonable results; that is, the functional dependence is quite good.

For $t \leq 0$ Eq. (6.10) becomes Eq. (6.14), since the diode is at dc steady state and the stored charge is not changing with time,

$$\frac{dQ_p}{dt} = 0 = I_F - \frac{Q_p(0)}{\tau_p} \tag{6.14a}$$

Because the charge cannot change instantaneously, $Q_p(0^-) = Q_p(0^+)$ and solving for $Q_p(0^+)$ from Eq. (6.14a) yields

$$Q_p(0^-) = Q_p(0^+) = I_F \tau_p \tag{6.14b}$$

Substituting $Q_p(0^+)$ from Eq. (6.14b) into Eq. (6.13) yields Eq. (6.15).

$$\boxed{t_s \cong \tau_p \ln\left[1 + \frac{I_F}{I_R}\right]} \tag{6.15}$$

An examination of Eq. (6.15) shows that indeed our qualitative arguments are valid. In particular t_s will decrease if τ_p and I_F are decreased, which is the same effect as decreasing $Q_p(0^+)$. Note that t_s decreases if I_R is increased; this is identical to increasing the diffusion current to remove holes from the n-region faster.

A more detailed analysis of the charge storage in the p^+-n junction results in Eq. (6.16).

$$\text{erf}\left(\sqrt{\frac{t_s}{\tau_p}}\right) = \frac{1}{1 + \dfrac{I_R}{I_F}} \tag{6.16}$$

Figure 6.5 illustrates the difference between Eqs. (6.16) and (6.15), showing the conservative estimate for t_s obtained by Eq. (6.15). It should also be noted that the experimental measurement of t_s is used to obtain values of τ_p (or τ_n in the case of a p-n^+ device).

Figure 6.6 illustrates the minority carrier concentration in the time region after t_s until t_{rr} and on to infinity. Note that the slope at $x = 0$ is changing, decreasing, as v_A goes negative. As the slope decreases, the hole current magnitude decreases (less negative) and the current approaches the reverse saturation current $(-I_0)$ as $t \to \infty$. When $|i(t)|$ reaches 10% of $|I_R|$ we define the reverse recovery time t_{rr}.

SEE EXERCISE 6.1 – APPENDIX A

6.3   THE TURN-ON TRANSIENT

The forward turn-on transient occurs when the diode is pulsed from reverse bias into a forward current conducting state. This can be accomplished with a current pulse, a voltage pulse, or a mixture of the two pulses. Because of its simplicity and direct con-

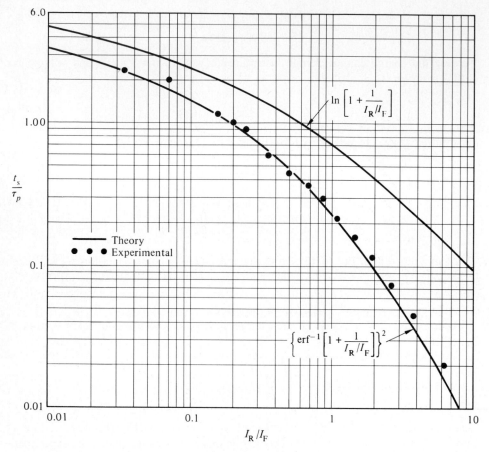

Fig. 6.5 t_s dependence.

nection to many practical circuits, we will present here only the current pulsed case.

When the diode current is pulsed instantaneously from $-I_0$ to some constant forward current I_F, the voltage response, $v_A(t)$, changes from some negative value at $t = 0$ to V_A at $t = \infty$. The first stage of the response, from $t = 0^+$ to where $v_A = 0$, occurs very rapidly, in about the dielectric relaxation time of the semiconductor ($\cong 10^{-10}$ sec or less). The response is very quick because the majority carriers are moving to narrow the deple- tion region back toward its thermal equilibrium value ($V_A = 0$). Electrons in the n-region and holes in the p-region neutralize the donor and acceptor ions.

Having disposed of negative bias to $v_A = 0$, let us next consider the case where the diode is pulsed with a current step from $i = 0$ to I_F at $t = 0$ as depicted by Fig. 6.7(a). The voltage response is shown in Fig. 6.7(b), while Fig. 6.8 illustrates how the minor- ity carriers change until steady state is attained. The current step injects holes into the

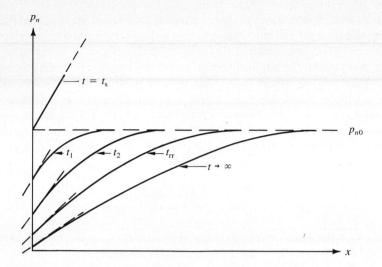

Fig. 6.6 Minority carrier depletion for $t_s < t < \infty$, at large magnification of vertical scale; the dashed line indicates the slope at $x = 0$.

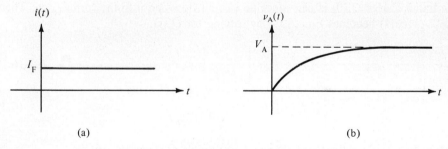

(a)   (b)

Fig. 6.7 Turn-on transient: (a) current; (b) voltage.

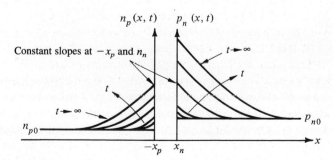

Fig. 6.8 Carrier concentrations during the turn-on transient.

n-region at a constant rate. Since the hole current at the edge of the depletion region (x_n) is constant, the slope of the hole concentration is constant. Similarly, the slope of the electron concentration is constant at $-x_p$ as illustrated in Fig. 6.8. Note that as $p_n(x_n, t)$ and $n_p(-x_p, t)$ increase, the terminal voltage $v_A(t)$ must increase according to Eqs. (6.4) and (6.5). Therefore we expect the voltage to increase from zero to some final value necessary to support the steady-state current I_F as illustrated in Fig. 6.7(b).

The analytical solution for the turn-on transient is simpler if we again assume a p^+-n junction where the total current is essentially the hole current injected at x_n. Applying Eq. (6.10) to the problem, the hole charge increases due to the current step, with recombination using up some of the charge. For $t \geq 0$, $i(t) = I_F$ and

$$\frac{dQ_p(t)}{dt} = I_F - \frac{Q_p(t)}{\tau_p}$$

The rate at which the charge builds up is due to the current I_F minus that lost to recombination. To solve for $Q_p(t)$ let's use a Laplace transformation.

$$sQ_p(s) - Q_p(0) = \frac{I_F}{s} - \frac{Q_p(s)}{\tau_p} \tag{6.17}$$

The initial charge $Q_p(0)$ is zero since no current flows prior to the current step. Therefore Eq. (6.17) becomes Eq. (6.19) by solving for $Q_p(s)$.

$$Q_p(s)\left[s + \frac{1}{\tau_p}\right] = \frac{I_F}{s} \tag{6.18}$$

$$Q_p(s) = \frac{I_F}{s(s + 1/\tau_p)} \tag{6.19}$$

The inverse Laplace transform of Eq. (6.19) yields $Q_p(t)$ as Eq. (6.20).

$$Q_p(t) = \tau_p I_F(1 - e^{-t/\tau_p}) \tag{6.20}$$

that is,

$$Q_p(\infty) = \tau_p I_F$$

from Eq. (6.20). Obviously the hole charge increases as the junction charges up to some final value.

To obtain an estimate of the voltage response shown in Fig. 6.7(b), let's assume that the hole distribution at each instant of time in Fig. 6.8 can be approximated by an exponential of the type

$$p_n(x',t) = p_{n0}e^{qv_A(t)/kT}e^{-x'/L_P} \tag{6.21a}$$

The total excess-hole charge is obtained by multiplying Eq. (6.21a) by qA and integrating over the entire bulk n-region, yielding Eq. (6.21b) as an approximation to $Q_p(t)$.

$$Q_p(t) = \tau_p I_F(1 - e^{-t/\tau_p}) \cong qAp_{n0}L_P(e^{qv_A(t)/kT} - 1) \tag{6.21b}$$

The voltage response can then be solved from Eq. (6.21b) as

$$v_A(t) \cong \frac{kT}{q} \ln\left[1 + \frac{\tau_p I_F}{qAp_{n0}L_P}(1 - e^{-t/\tau_p})\right] \tag{6.22a}$$

$$\boxed{v_A(t) \cong \frac{kT}{q} \ln\left[1 + \frac{I_F}{I_0}(1 - e^{-t/\tau_p})\right]} \tag{6.22b}$$

Several observations can be made about the turn-on transient by inspection of Eq. (6.22). A smaller value of I_F and/or a smaller value of τ_p will give a faster turn-on time. Because $Q_p(\infty) = I_F \tau_p$, these results can be generalized as demonstrating that the smaller the amount of final stored charge, the quicker the device turns on. Again the amount of charge transfer controls the switching speed.

SEE EXERCISE 6.2 – APPENDIX A

6.4 SUMMARY

The large-signal pulsed response of the p-n junction was considered in terms of the turn-off and turn-on transient responses. When switched off from a conducting state, the current response was characterized by the reverse recovery time, t_{rr}. The longest component of t_{rr} is the storage time t_s. A reduction in the forward current and/or minority carrier lifetime reduces t_s. If the magnitude of the reverse current is made larger, t_s is made smaller. Any means of reducing the minority carrier charge storage before switching reduces the reverse recovery time.

When switched on from the off state, the diode is controlled by the time necessary to build up the minority carrier charge in the bulk n- and p-regions. We considered the case of a current step from zero to I_F resulting in a buildup of the junction voltage from zero to a final value of V_A. Again, any means of reducing the minority carrier charge storage will reduce the switching time. In particular, if I_F and/or τ_p and τ_n are reduced, the device will respond more quickly. The original interest in diode transients was motivated by the desire to minimize the reverse recovery time and thereby reduce the switching time of diode logic circuits.

PROBLEMS

6.1 A p^+-n step junction has in Fig. 6.1 $V_F = 10$ volts, $V_R = 12$ volts, $R_F = 10$ kΩ, and $\tau_p = 5$ μs. Assume that $Q_p(t_s) = 0$ and calculate:

(a) t_s if $R_R = 5$ kΩ;

(b) t_s if $R_R = 1$ kΩ;

(c) a value of R_R to make $t_s = \tau_p/5$;

(d) a value of R_R to make $t_s = 2\tau_p$.

6.2 An abrupt p^+-n junction is conducting current in the forward direction of I_F and is then switched to off at $t = 0$ through a current source of

$$i_s = I_F e^{-t/\tau_p}$$

Derive an equation for $Q_p(t)$ and give a rough sketch of the result.

6.3 A p^+-n step junction is switched from a forward current source of I_F to zero at $t = 0$.

(a) Derive an equation for $Q_p(t)$ in terms of τ_p and I_F; sketch $p_n(x, t)$.

(b) If your plot of $p_n(x, t)$ is approximated as an exponential in "x," derive an equation for $v_A(t)$. Assume that in the range of interest the junction is well into forward bias.

(c) How might this experiment be used to measure τ_p? Explain.

6.4 A p^+-n step junction is forward biased at I_{F1} and pulsed to a constant current of I_{F2} at $t = 0$ as shown in Fig. P6.4.

(a) Sketch the hole carrier concentration as a function of t and x.

(b) Calculate $V_A (\infty) - V_A(0^-)$.

(c) Derive a formula for $Q_p(t)$.

6.5 If a p^+-n step junction is switched from $I = 0$ to 1 mA with a current pulse at $t = 0$, calculate the time necessary for the device to reach 90% of its final voltage. Let $\tau_p = 1.0$ μs and $I_0 = 10^{-15}$ A.

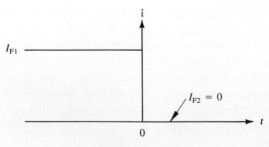

Fig. P6.3

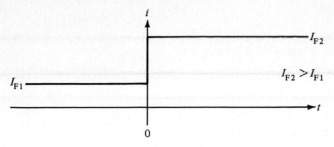

Fig. P6.4

7 / Metal – Semiconductor Contacts

7.1 INTRODUCTION

Metal – semiconductor contacts play a very important role in all solid state devices. They are the route from the semiconductor to the outside world, they interconnect devices within integrated circuits, and they can become a rectifying junction or be an effective ohmic contact. Here we present the basic concepts of the Schottky barrier diode and the non-rectifying (ohmic) contact. First the energy bands for the metal and the semiconductor, separated from each other, are discussed in light of the work function of the metal and the electron affinity of the semiconductor. Placing the metal in intimate contact with the semiconductor and how this affects the thermal equilibrium energy band diagram is next presented along with the potential barriers for electrons. The other sections of the chapter present the ideal Schottky diode's V-I characteristics, its depletion capacitance, and finally the several deviations from the ideal device.

In some respects the Schottky barrier diode resembles the p^+-n step junction with the p^+ silicon replaced by the metal. The n-type region has numerous electrostatic similarities. Many of the concepts developed in Chapters 2 and 3 will apply to this device. Metal – silicon contacts used as "ohmic" contacts finish the chapter.

7.2 METAL – SEMICONDUCTOR THERMAL EQUILIBRIUM ENERGY BAND DIAGRAMS

The energy band diagrams for a metal and for a semiconductor are illustrated in Fig. 7.1(a) and (b), respectively. They are separated by a nearly infinite distance so they do not interact with each other and remain isolated. An energy level, E_0, is new to the energy band diagrams. This energy is the *vacuum energy level,* and represents the energy at which an electron can be "free" of the material. By that we mean free to wander off into the surrounding vacuum and be emitted away from the solid. The vacuum energy (E_0) is a convenient energy reference level from which to compare two dissimilar materials.

The *work function* (Φ) of a material is the energy difference from the vacuum energy (E_0) to the Fermi energy level (E_F). For the metal of Fig. 7.1(a), Φ_M is the work

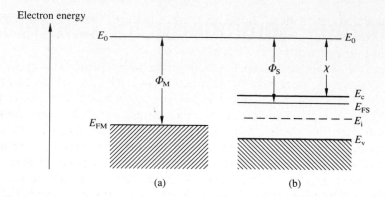

Fig. 7.1   Energy band diagrams for isolated (a) metal and (b) n-type semiconductor.

function and its value depends on the type of metal, i.e., whether it is gold, aluminum, platinum, etc. The value of Φ_M is a fundamental property of the particular metal. Note that the units of Φ are in "electron volts," an energy unit.

The work function for the semiconductor, Φ_S, is shown in Fig. 7.1(b). It depends on the semiconductor doping because the position of E_{FS} depends on the type of doping and the concentration of N_A or N_D. However, the energy difference between the vacuum level and the conduction band edge is a fundamental property of the semiconductor, independent of doping. Hence the **electron affinity** energy (χ) for the semiconductor is defined as $(E_0 - E_c) = \chi$. For silicon it has a value of 4.05 eV. Therefore the work function for a semiconductor is the energy difference between E_0 and the Fermi energy, E_{FS}, as given by Eq. (7.1).

$$\Phi_S = \chi + (E_c - E_{FS})_{bulk} \quad (eV) \tag{7.1}$$

Remember that $E_c - E_F$ is in the bulk region and is dependent on the doping concentration of the semiconductor.

Consider the case, as shown in Fig. 7.1, where $\Phi_M > \Phi_S$ for the metal and n-type semiconductor. For the infinitely separated materials shown, the Fermi level in the metal (E_{FM}) is less than the Fermi level in the semiconductor (E_{FS}). Therefore the average energy of electrons in the semiconductor is greater than the average energy of those in the metal. When we bring the two materials into intimate (perfect) contact, the difference in the average electron energy can be expected to transfer electrons from the semiconductor to the metal until the average electron energies are equal. This can be accomplished by waiting for the system to return to thermal equilibrium.

A transfer of electrons from the n-type semiconductor to the metal leaves the surface of the semiconductor depleted of electrons and leaves behind some positive donor ions. Having received electrons, we expect the metal to be negatively charged with respect to the semiconductor. Note how similar the n-side is to the p^+-n junction, a depletion region with positive (fixed) charge density. The metal has a surface charge that is negative by the same amount the semiconductor is positive.

To construct a valid energy band diagram for the metal–semiconductor (M–S) contact it is necessary to emphasize several major points. The first point is that the work functions for the metal and electron affinity of the semiconductor are fundamental properties of the two materials and therefore cannot change just because we brought them into contact. Hence the vacuum level, E_0, must be a continuous function with respect to E_{FM} and to E_c. The second point, as was the case for the p-n junction at thermal equilibrium, is that the Fermi energy level is a constant. With these points in mind, push the two energy band diagrams of Fig. 7.1(a) and (b) together horizontally as illustrated in Fig. 7.2(a) and (b). They come into intimate contact at $t = 0$ as shown in Fig. 7.2(b); i.e., they are not yet at thermal equilibrium. Some electrons, in this particular case, leave the semiconductor and go to the metal. The thermal equilibrium diagram of Fig. 7.2(c) is obtained by pulling down the right-hand side of Fig. 7.2(b) until $E_{FM} = E_{FS}$ with the surface of the semiconductor ($x = 0$) fixed in place. Here we consider the ideal case where no surface effects are included. Pulling the right-hand side down represents a loss of negative charge (electrons) from the n-type semiconductor. Also, remember to keep $E_0(x = 0^-)$ equal to $E_0(x = 0^+)$ and that χ is a constant. Deep into the bulk region of the semiconductor the energy band becomes constant.

Figure 7.2 illustrates several interesting points. One is that the barrier for an electron with energy $E = E_{FM}$ sees a potential barrier towards the semiconductor of Φ_B, which is the difference between Φ_M and χ. An electron deep in the semiconductor bulk at $E = E_c$ sees a potential barrier towards the metal which is the difference in the two original Fermi energy levels (before contact). Note that the energy band bending in the semiconductor is quite similar to the p^+-n junction. It is important to remember that Fig. 7.2 is for the ideal M–S contact where effects at their interface are not included.

7.3 IDEAL SCHOTTKY BARRIER DIODE ELECTROSTATICS

The metal–semiconductor (M–S) contact in which $\Phi_M > \Phi_S$ is called a Schottky barrier diode and there is nonlinear current flow through the device. Because the current flows easily in one direction but not in the other, it is a rectifying contact. In this section we consider the charge density (ρ), electric field ($\mathscr{E}$), and potential (V) of this type of diode.

Figure 7.3(a) illustrates the thermal equilibrium energy band diagram and the two energy barriers for electrons Φ_B and qV_{bi}. Note that they are in energy units of electron volts while V_{bi} is in units of volts. Invoke the "depletion approximation," as was the case for the ideal p-n junction. As a result of the approximation the charge density (ρ), as illustrated in Fig. 7.3(b), is "squared off" near the depletion region edge, x_n, in the bulk semiconductor. Since the electrons have left the semiconductor, which is n-type, the charge density is constant for a uniformly doped semiconductor. Note that near the surface of the semiconductor it is depleted (of electrons) as shown in Fig. 7.3(a). At

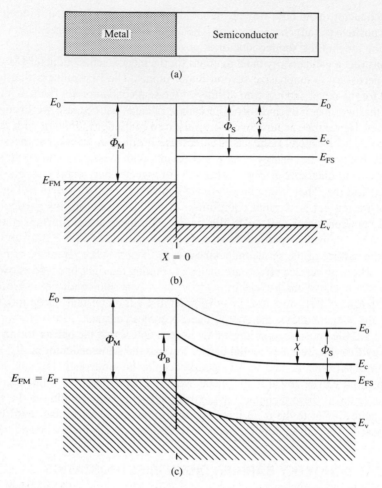

Fig. 7.2 M–S Schottky barrier: (a) physical; (b) nonthermal equilibrium at instant of contact, $t = 0$; (c) thermal equilibrium.

the surface of the metal is a surface charge of negative electrons. Because of charge neutrality the two charges can be equated as

$$Q_S = qN_D x_n = -Q_M \quad (\text{coul/cm}^2) \tag{7.2}$$

The electric field ($\mathscr{E}$) is determined qualitatively from the slope of the energy band diagram. In this case at the surface ($x = 0$) it is negative. Progressing into the semiconductor the field becomes less negative until x_n, where it becomes zero. Applying a form of Eq. (2.1), the electric field is derived between $x = 0$ and $x = x_n$ as

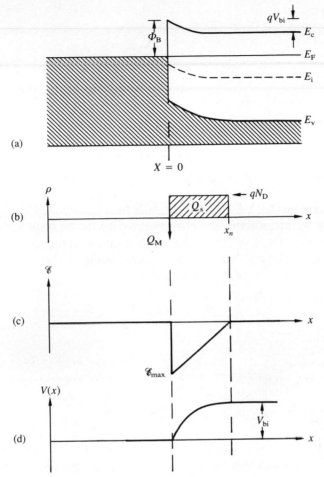

Fig. 7.3   Schottky barrier in thermal equilibrium: (a) energy band; (b) charge density; (c) electric field; (d) potential.

$$\mathcal{E}(x) = \frac{1}{K_s \varepsilon_0} \int q N_D^+ \, dx + A_1 = \frac{q N_D x}{K_s \varepsilon_0} + A_1 \qquad (7.3)$$

Since $\mathcal{E}(x_n) = 0$,

$$\mathcal{E}(x_n) = \frac{q N_D x_n}{K_s \varepsilon_0} + A_1 \equiv 0$$

Therefore,

$$A_1 = -\frac{qN_D x_n}{K_S \varepsilon_0}$$

and then

$$\mathscr{E}(x) = -\frac{qN_D}{K_S \varepsilon_0}(x_n - x) \tag{7.4}$$

which is plotted in Fig. 7.3(c). Note that Eq. (7.3) is identical to Eq. (2.22) for the same *n*-region of the *p-n* step junction.

The potential function, $V(x)$, can be obtained from the shape of $E_c(x)$, as was done with the *p-n* diode, by reflecting about the horizontal axis. Qualitatively, the potential difference across the semiconductor is the amount by which the energy band diagram was shifted to reach thermal equilibrium as shown in Fig. 7.3(a), i.e., the difference between the work functions of the two materials.

$$qV_{bi} = \Phi_M - \Phi_S \tag{7.5a}$$

From Fig. 7.2(b)

$$qV_{bi} = \Phi_M - \chi - (E_c - E_F)_{bulk} \tag{7.5b}$$

Therefore $V(x_n) = V_{bi}$ if we select $V(x = 0) = 0$ as shown in Fig. 7.3(d).

Quantitatively, the potential function is derived by using a form of Eq. (2.4):

$$V(x) = -\int \mathscr{E}(x)\,dx + A_2$$

From Eq. (7.4),

$$V(x) = -\int \left(-\frac{qN_D}{K_S \varepsilon_0}\right)(x_n - x)\,dx + A_2 = \frac{qN_D}{K_S \varepsilon_0}x_n x - \frac{qN_D}{K_S \varepsilon_0}\left(\frac{x^2}{2}\right) + A_2$$

Since we have chosen a zero potential for the metal,

$$V(x_n) = V_{bi} = \frac{qN_D x_n x_n}{K_S \varepsilon_0} - \frac{qN_D}{K_S \varepsilon_0}\frac{x_n^2}{2} + A_2$$

Therefore

$$A_2 = V_{bi} - \frac{qN_D}{K_S\varepsilon_0}\frac{x_n^2}{2}$$

Now we can write the potential function as Eq. (7.6).

$$V(x) = V_{bi} - \frac{qN_D}{2K_S\varepsilon_0}(x_n - x)^2 \qquad (7.6)$$

Note that this equation is a parabola and that from the boundary condition $V(x = 0) = 0$ then

$$V_{bi} = \frac{qN_D}{2K_S\varepsilon_0}x_n^2$$

Inverting and solving for x_n yields

$$x_n = \sqrt{\frac{2V_{bi}K_S\varepsilon_0}{qN_D}} \qquad (7.7)$$

where V_{bi} is evaluated from Eq. (7.5b).

7.3.1 Applied Bias

A voltage is applied from the metal to the semiconductor, V_A, in much the same manner as the p-n junction. The negative applied potential on the n-type semiconductor causes that side of the energy band diagram to shift upward (or the metal side to shift downward) and reduce the barrier height for electrons going from the semiconductor toward the metal from qV_{bi} to $q(V_{bi} - V_A)$, as shown in Fig. 7.4(a) for $V_A > 0$. Similar to the arguments used for the p-n junction, the value of x_n is reduced, as is the magnitude of the electric field and potential. We can now write Eq. (7.7) as

$$x_n = \sqrt{\frac{2K_S\varepsilon_0}{qN_D}(V_{bi} - V_A)} \qquad (7.8)$$

When $V_A > 0$, the condition is called forward bias, since now the electrons can cross from the semiconductor to the metal more easily due to the lower barrier. Note that these

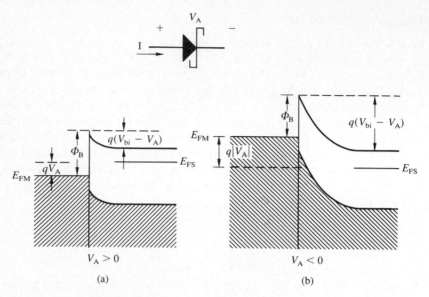

Fig. 7.4 Schottky diode with applied voltages: (a) forward bias; (b) reverse bias.

are the majority carriers in the n-type semiconductor. As V_A increases, x_n decreases and the maximum electric field is less negative.

At $V_A < 0$ the diode is reverse biased and now the semiconductor is pulled down in the electron energy band diagram, as illustrated in Fig. 7.4(b). Here the barrier height for electrons in the semiconductor has increased by $|qV_A|$. From Eq. (7.8) the depletion width, x_n, is increased while from Eq. (7.4) the electric field is more negative. Now the number of electrons flowing from the semiconductor to the metal is very small due to the large barrier.

SEE EXERCISE 7.1 – APPENDIX A

7.4 IDEAL SCHOTTKY BARRIER *I–V* CHARACTERISTICS

The Schottky barrier diode has many applications in modern integrated circuits. One such application is as a shunting diode for TTL logic. A "Schottky clamp" placed in parallel with the p-n junction causes the switching time to be reduced by about a factor of ten. This application makes good use of the rectifying I–V characteristics of the Schottky device and that it has little or no minority carrier charge storage. Problem 7.4 is an example of the diode shunt. The bias dependence of the current is first deduced from qualitative arguments. This provides fundamental insight into how the diode operates.

7.4.1 Thermal Equilibrium

First consider the electron flux of carriers as they attempt to go from the metal to the semiconductor. An electron at E_{FM} sees an energy barrier of Φ_{B}. Therefore it must have at least that energy and a velocity that is x-directed in order to reach the n-type semiconductor. In Fig. 7.4(a) and (b) we see that this barrier (Φ_{B}) **does not change** with applied bias, V_{A}. We also know that the electrons are distributed in energy by the Fermi function, as discussed in Chapter 2 of Volume I in this series. Therefore only some of the electrons have energies greater than Φ_{B} and can go from the metal into the semiconductor. Figure 7.5 illustrates the electrons that can surmount the barrier Φ_{B} and traverse the depletion region into the bulk semiconductor. We call this current density component $J_{\text{M–S}}$. The Fermi function is the probability that an electron can have an energy, E. It can be approximated as

$$f(E) = \frac{1}{1 + e^{(E-E_{\text{F}})/kT}} \cong e^{-(E-E_{\text{F}})/kT} \tag{7.9}$$

for $E > 3kT + E_{\text{F}}$. Then, for $\Phi_{\text{B}} > 3kT$ the number of electrons with energies above the barrier decreases approximately exponentially with increasing electron energy.

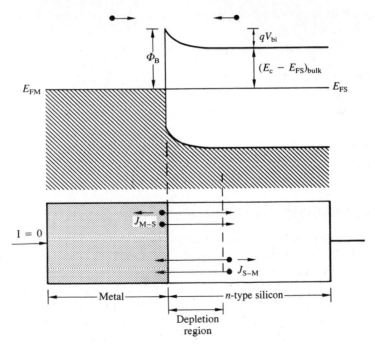

Fig. 7.5 Schottky barrier diode at thermal equilibrium, $V_{\text{A}} = 0$.

Hence the larger Φ_B, the fewer electrons can cross into the semiconductor, and we expect the carrier flux to decrease exponentially as the barrier height increases.

Under equilibrium conditions, $V_A = 0$, electrons at the edge of the conduction band in the bulk semiconductor see a potential barrier of qV_{bi}. The number of electrons with energies greater than $E_c + qV_{bi}$ also decreases exponentially with increasing energy for a nondegenerate material or where $qV_{bi} + (E_c - E_{FS}) > 3kT$. These bulk semiconductor electrons can travel into the metal since they have enough energy to get over the barrier. Again, the larger qV_{bi}, the smaller the number that can make the transition. Figure 7.5 shows these electrons as the current density component J_{S-M}.

The Schottky diode is in thermal equilibrium; therefore the number of electrons traversing from right to left (J_{S-M}) must equal the number going from left to right (J_{M-S}). Electrons can't just build up at one end of the device and still be in thermal equilibrium. Let's estimate the currents using the concept that the current is proportional to the number of electrons at the surface of the semiconductor that have energies larger than the barrier. First let's go back to Volume I for the number of electrons in the conduction band, repeated here as Eq. (7.10).

$$n = N_C e^{-(E_c - E_F)/kT} \tag{7.10}$$

The above equation represents the number of electrons at the surface if we let $n(x = 0) = n_S$. Then the energy difference from E_F to the conduction band is Φ_B. Therefore

$$n_S = N_C e^{-(\Phi_B)/kT} \tag{7.11}$$

Also from Fig. 7.5, $\Phi_B = qV_{bi} + (E_c - E_{FS})_{bulk}$. Substituting this into Eq. (7.11) yields

$$n_S = N_C e^{-qV_{bi}/kT} e^{-(E_c - E_{FS})_{bulk}/kT} = N_C e^{-(E_c - E_{FS})_{bulk}/kT} e^{-qV_{bi}/kT} \tag{7.12}$$

Note that the term

$$N_C e^{(E_c - E_{FS})_{bulk}/kT} = n \tag{7.13}$$

is the bulk majority carrier concentration, $n \cong N_D$. We can now write the electrons at the surface as

$$n_S = N_D e^{-qV_{bi}/kT} = N_C e^{-\Phi_B/kT} \tag{7.14}$$

Now assuming the current J_{M-S} is proportional to n_S and therefore

$$J_{M-S} = -K_1 N_D e^{-qV_{bi}/kT} \tag{7.15a}$$

or from Eq. (7.14),

$$J_{M-S} = -K_1 N_C e^{-\Phi_B/kT} \tag{7.15b}$$

Where K_1 is a constant of proportionality.

Electrons in the semiconductor must have a flux equal and opposite since the device is in thermal equilibrium. Therefore,

$$|J_{M-S}| = |J_{S-M}|, \qquad J_{S-M} = K_1 N_D e^{-qV_{bi}/kT} \tag{7.16a}$$

or as

$$J_{S-M} = K_1 N_C e^{-\Phi_B/kT} \tag{7.16b}$$

which are in terms of the two barriers. Note that for this device type we are considering only the majority carrier electrons in the semiconductor. The holes are the minority carriers and contribute very little to the total current. From Fig. 7.5 it is evident that no barrier exists for the thermally generated holes to leave the semiconductor and go to the metal and recombine. Later we discuss this as an "ohmic" contact for the holes.

7.4.2 Forward Bias, $V_A > 0$

The applied voltage changes the potential barrier for electrons in the semiconductor but not for electrons in the metal, as shown in Fig. 7.6. Therefore J_{M-S} remains the same as Eq. (7.15a) or (7.15b) while J_{S-M} increases rapidly with increasing V_A. With V_A applied the barrier for electrons in the semiconductor becomes $q(V_{bi} - V_A)$ and Eq. (7.16a) can be written as

$$J_{S-M} = K_1 N_D e^{-q(V_{bi}-V_A)kT} \tag{7.17}$$

The total current is the sum of J_{S-M} and J_{M-S}, which from Eq. (7.15a) is

$$J = K_1 N_D e^{-q(V_{bi})/kT} e^{qV_A/kT} - K_1 N_D e^{-qV_{bi}/kT}$$

Factoring gives

$$J = K_1 N_D e^{-qV_{bi}/kT} [e^{qV_A/kT} - 1]$$

However, comparing Eqs. (7.15a) and (7.15b) note that

$$N_D e^{-qV_{bi}/kT} = N_C e^{-\Phi_B/kT}$$

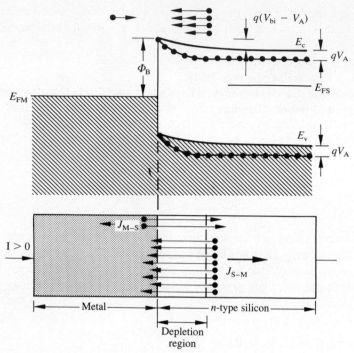

Fig. 7.6   Schottky barrier diode: forward biased, $V_A > 0$ (————); thermal equilibrium (•••••).

and since Φ_B and V_{bi} do not change with V_A, the current density equation can be written as

$$J = K_1 N_C e^{-\Phi_B/kT}[e^{qV_A/kT} - 1] \qquad (7.18a)$$

The reverse saturation current for the Schottky barrier diode is defined as

$$J_0 = K_1 N_C e^{-\Phi_B/kT} = K e^{-\Phi_B/kT} \qquad (7.18b)$$

This current is the limit of the reverse bias current, the electrons in the metal having energies greater than $E_{FM} + \Phi_B$. Equation (7.18a) can then be written as

$$J = J_0[e^{qV_A/kT} - 1] \qquad (7.19)$$

The net current under forward bias results from the large increase in electrons going from the semiconductor to the metal, as shown in Fig. 7.6 due to the lower barrier.

7.4.3 Reverse Bias, $V_A < 0$

Figure 7.7 illustrates the electron flux when reverse biased, $V_A < 0$. The potential barrier for electrons in the semiconductor has increased and the result is a drastic decrease in the number of electrons that can get into the metal. Electrons in the metal, however, see the same barrier as before, Φ_B. It did not change with reverse bias.Therefore the J_{M-S} current remains the same. Equations (7.19) and (7.18a) become

$$J \cong -J_0 = -Ke^{-\Phi_B/kT}$$

when $V_A \ll 0$, which also agrees with Eq. (7.15b) for the metal-to-semiconductor electron current.

In deriving an equation for the ideal Schottky barrier diode we made several simplifying assumptions that are not exact. A more rigorous solution to the problem shows that the constant K in Eq. (7.18b) has a $\sqrt{(V_{bi} - V_A)}$ dependence on V_A. Since this is a

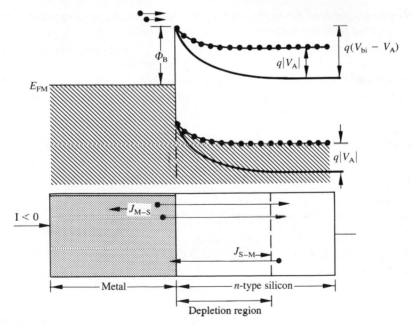

Fig. 7.7   Schottky barrier diode in reverse bias, $V_A < 0$ (———); in thermal equilibrium (————).

slowly varying function as compared to the exponential, we can use Eq. (7.20) as the approximate equation

$$J = J_0(e^{qV_A/nkT} - 1) \qquad\qquad (7.20)$$

where the ideality factor (n) typically ranges from 1.00 to 1.10 in practical devices. The current J_0 reflects the change in the barrier height.

SEE EXERCISE 7.2 – APPENDIX A

7.5 DEPLETION CAPACITANCE

The depletion region width in the n-type semiconductor (x_n) changes with V_A as discussed in Section 7.3. If in addition to the dc bias a small-signal sine wave is applied to the Schottky barrier, $v_a = V_m \sin \omega t$, the total applied junction voltage is

$$v_A = V_A + v_a$$

The charge increment at the depletion region edge must change to accommodate v_a. Because the change in the depletion width is made by moving the majority carrier electrons, the process is very fast. Hence the metal, depletion region, and semiconductor layers form a capacitance very similar to the p-n depletion capacitance. That is, the metal to depletion region to bulk semiconductor is like a parallel plate capacitor with the depletion region acting as the insulator. Therefore the depletion capacitance, from Eq. (7.8), is

$$C = \frac{K_S \varepsilon_0 A}{x_n} = \frac{K_S \varepsilon_0 A}{\sqrt{\dfrac{2K_S \varepsilon_0}{qN_D}(V_{bi} - V_A)}} \quad \text{(F)} \qquad (7.21)$$

where A is the cross-sectional area (cm^2). Dividing by $K_S \varepsilon_0$ yields

$$C = \frac{A}{\sqrt{\dfrac{2(V_{bi} - V_A)}{qN_D K_S \varepsilon_0}}} \qquad\qquad (7.22)$$

Using the zero bias ($V_A = 0$) capacitance definition (C_0) and factoring it out of Eq. (7.22) yields

$$C = \frac{C_0}{\left[1 - \dfrac{V_A}{V_{bi}}\right]^{1/2}}$$

(7.23)

where

$$C_0 = \frac{A}{\sqrt{\dfrac{2V_{bi}}{qN_D K_S \varepsilon_0}}} \quad \text{(F)}$$

(7.24)

Equation (7.22) can be inverted to obtain $1/C^2$:

$$\frac{1}{C^2} = \frac{2(V_{bi} - V_A)}{qN_D K_S \varepsilon_0 A^2}$$

(7.25)

which is plotted in Fig. 7.8. Note that the projection to where $1/C^2 = 0$ yields V_{bi} and that the slope is proportional to N_D^{-1} when the semiconductor is uniformly doped. This property can be used to measure the doping density of silicon wafers.

The Schottky barrier diode has majority carrier electrons responding to v_a in n-type silicon. Therefore the depletion capacitance is independent of frequency up to very high frequencies. In addition, the device has almost no diffusion capacitance as compared to the p-n junction. This is primarily due to the little, if any, minority carrier storage in the

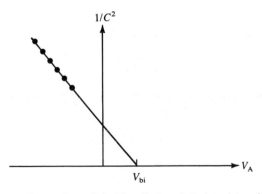

Fig. 7.8 Depletion capacitance for a Schottky diode, plotted to determine doping and barrier height.

semiconductor. Remember that in the *p-n* junction, once the majority carriers cross the depletion region of the junction they become minority carriers. As a result minority carriers are stored in both the *n*-type and *p*-type bulk regions. Here electrons are majority carriers on both sides of the junction. The result is a diode and less total capacitance when forward biased. It therefore can operate at much higher frequencies than the *p-n* junction.

Switching speeds for the Schottky barrier diode are very short. Because there is little, if any, storage of minority carrier holes in the *n*-type semiconductor, the diode's turn-on and turn-off times are limited only by movement of the majority carriers. Their movement is very quick, being limited by the dielectric relaxation time of the material.

7.6 NON-IDEALITIES FOR SCHOTTKY BARRIERS

The most obvious deviation from ideal is the series resistance associated with the bulk *n*-type semiconductor. As with the *p-n* junction it can, at large forward currents, cause an additional voltage drop across the diode. The result is a deviation from Eq. (7.19) or Eq. (7.20).

A second deviation from ideal is that Φ_B is not entirely independent of V_A. An advanced treatment will show that due to "Schottky barrier lowering" from the electric field at the metal – semiconductor interface, Φ_B is slightly lower than as presented. This field effect shows a fourth-root dependence on $(V_{bi} - V_A)$, and hence not a large change in Φ_B with reverse voltage. However, from Eq. (7.18b) a lowering of Φ_B results in a much larger reverse saturation current that increases with larger reverse bias. This effect is often observed in many practical devices that have relatively low values of the barrier, Φ_B.

The reverse saturation current in a Schottky barrier diode with a large Φ_B is most dramatically affected by generation of electrons and holes in the depletion region. As in the *p-n* junction, in silicon, the generation current can increase the reverse current at room temperature. Similar to Eq. (4.11), and as discussed in Volume I, Chapter 4, we can write the generation current in reverse bias as

$$J_{R-G} \cong -q\frac{n_i x_n}{2\tau_0} \tag{7.26}$$

Note that x_n will increase as the square root of reverse bias.

The last deviation from ideal we will discuss is the metal – silicon interface which was assumed to be perfect, which of course is not the case in real devices. First, our representation of the semiconductor assumed that it was an infinite number of atomic distances thick and that all the silicon bonds were complete in the diamond lattice. At the edge of the silicon crystal, near the metal, not all the covalent atomic bonds can be satisfied by other silicon atoms. Therefore the surface has crystal defects. An unsatisfied silicon bond results in an energy state in the band gap at the surface of the semiconductor. Because the energy state (level) is only at the surface between the metal and the semiconductor, it is called an "interface state."

Interface states can also occur due to foreign (contaminant) atoms at the surface of the silicon and/or due to surface roughness (crystal damage). For these and other reasons the interface is not ideal and defects may alter the height of the electrical barrier, Φ_B. Volume IV of this series will discuss these interface states in more detail. Although they are not completely understood, some general comments can be made about them. First, there is a range of energy levels in the band gap which are a result of interface states. Second, the number density of these levels is not uniform in energy. They are also affected by the metal deposition conditions. Each surface preparation technique should be considered, as each may give different results.

The major effect of the interface states is that they can trap charge and therefore change the value of Φ_B. Depending upon the type of interface state and the semiconductor doping, the barrier may be increased or decreased. To some extent the surface states determine the barrier height (or at least greatly influence it) in most practical applications. For n-type silicon with an aluminum metal, Φ_B is typically about 0.69 eV, while the platinum value is 0.85 eV.

Interface states will affect the barrier height, which in turn affects the $I - V$ characteristic values, but it does not give rise to any major changes in the conceptual operation of the Schottky barrier diode. However, a larger or smaller value of Φ_B will result in a large change in the reverse saturation current as predicted by Eq. (7.18b). In the forward-biased mode the current value at a given voltage is affected by the barrier height. Generally, Eq. (7.20) takes these effects into account and represents the experimental data very well over many decades of current in the forward-biased mode.

One parting comment: It should be noted that certain metals on p-type semiconductors can form Schottky barriers for holes. The operation can be explained in a similar manner.

7.7 OHMIC METAL–SILICON CONTACTS

Metal-to-semiconductor ohmic contacts to solid state devices and integrated circuit components are very important to practical applications. A perfect ohmic contact would have no voltage drop across the metal semiconductor interface for current flowing in either direction. A good ohmic contact would have very little voltage drop even at large current levels, and that voltage drop would be the same for both directions of current flow. We consider here two cases: (1) tunneling barrier, typically a metal on degenerately doped (n^+) n-type silicon where $\Phi_M > \Phi_S$; and (2) metal semiconductors where $\Phi_M < \Phi_S$ in n-type silicon.

7.7.1 Tunnel Contact

A very common ohmic contact to n-type silicon in integrated circuit technologies is the aluminum-to-N^+-silicon junction. The degenerately doped (very large $N_D \sim 10^{19}$ to $10^{20}/\text{cm}^3$) silicon has its Fermi level very near E_c. If $\Phi_M > \Phi_S$ we can draw the energy band diagram using the same methods as Fig. (7.2). In this case, however, N_D is very large so that from Eq. (7.7) the depletion into the semiconductor (x_n) is very small, leaving a very narrow barrier. Chapter 4 discussed tunneling between n-type and p-type semiconductors with respect to Zener breakdown voltages. Tunneling through the potential barrier can occur if it is very narrow, there are a large number of electrons on one side available to tunnel, and the other side of the barrier has a large number of empty states to tunnel into.

At thermal equilibrium the number of electrons tunneling through the barrier from the metal to the semiconductor is exactly balanced by the number tunneling from the semiconductor to the metal. Figure 7.9(a) shows the condition where a positive voltage is applied to the n-type material and the semiconductor has been lowered in the energy band diagram. Here the electrons tunnel easily from the metal into the relatively free states in the semiconductor. The more the semiconductor is lowered in the diagram, the larger the number of electrons to tunnel; hence the electron current increases very rapidly, requiring only a small voltage. Also note that the tunneling current from the semiconductor to the metal has been suppressed by fewer empty states to tunnel into in the metal.

Figure 7.9(b) illustrates the case where a negative potential was applied to the semiconductor. Now electrons can easily tunnel through the potential barrier from the

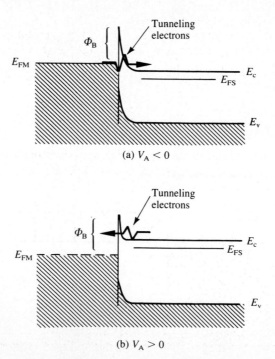

(a) $V_A < 0$

(b) $V_A > 0$

Fig. 7.9 Metal-to-semiconductor tunneling barrier on N^+ silicon: (a) tunneling from the metal to the semiconductor; (b) tunneling from the semiconductor to the metal.

semiconductor into the metal. Here the availability of empty states in the metal increases rapidly with applied bias. Therefore the electron tunneling current becomes very large even for small applied voltages. Also note that because of E_G, where no states exist, the electron tunneling from the metal to the semiconductor is suppressed.

Metal Schottky tunnel barriers are possible on degenerate p-type semiconductors in which majority carrier holes flow to or from the metal via tunneling through the potential barrier. Arguments similar to those presented above apply to this type of contact.

The tunnel contact has majority carrier flow to or from the semiconductor depending upon the voltage polarity. As long as the currents are large for small applied voltages, they approach our ideal "ohmic contact." In reality they do not have a straight-line $I - V$ plot, but if the potential drop across them is small compared to the device voltage we call them ohmic.

7.7.2 Negative V_{bi} Contact

An "ohmic contact" can be made to an n-type semiconductor by using a metal such that $\Phi_M < \Phi_S$. The energy band diagram for this situation is constructed by pulling the metal down in energy until the two Fermi energies are equal ($E_{FM} = E_{FS}$). Figure 7.10 illustrates this case. When the metal is separated from the semiconductor, $E_{FM} > E_{FS}$ and therefore the metal has an average electron energy greater than that of the semiconductor. Upon bringing the two materials together in intimate contact, electrons flow from the metal to the n-type semiconductor. Note that at the semiconductor surface the electron concentration is larger than in the bulk, as illustrated in Fig. 7.10. To bring the system into thermal equilibrium, electrons have left the metal and are accumulated at the surface of the semiconductor. One may even think of this as a metal-N^+-N sandwich.

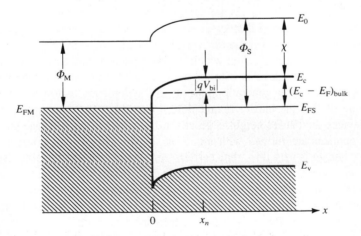

Fig. 7.10 Negative V_{bi}-type ohmic contact.

Inspection of Fig. 7.10 reveals that the surface charge in the semiconductor consists of "free," mobile electrons, which was not the case for the Schottky barrier diode. In that case it consisted of "fixed" positive donor ions and the surface was depleted. Remember that the closer E_c is to E_{FS}, the increase in electron concentration is exponential. Therefore the largest concentration of electrons is at the surface. Also note from the figure that electrons in the conduction band of the semiconductor see no potential barrier towards the metal.

Electrons in the metal have a small barrier to the conduction band, qV_{bi}, which can be calculated from Eq. (7.14) as a negative value, i.e., opposite polarity to the Schottky barrier diode. Another way is the use of the magnitude of Eq. (7.4) as

$$|V_{bi}| = |\chi + (E_c - E_{FS}) - \Phi_M| \tag{7.27}$$

It is interesting to note that for many metals and silicon, χ and Φ_M are very close in value. Therefore, with a heavily doped semiconductor the barrier is very small. The result is an excellent ohmic contact with very little voltage drop at large currents.

The discontinuity of the energy band at the surface of this type M–S contact can be calculated as $\Phi_M - \chi$. Since for aluminum and silicon this difference is about 0.05 eV, it is not as much a barrier as might be indicated by Fig. 7.10. Therefore the discontinuity has only a small effect on the electrical characteristics of the contact. In particular, there is almost no barrier and hence the electrons are quite easily moved from the metal to the semiconductor and vice versa.

7.8 SUMMARY

The metal contact to a semiconductor can result in an ideal Schottky barrier diode or an ohmic contact. When $\Phi_M > \Phi_S$ occurs on n-type silicon, a Schottky diode should result. If $\Phi_M < \Phi_S$ then an ohmic contact will occur. A practical device will have interface states that can affect or even determine the potential barrier, Φ_B.

The $I-V$ curve for the Schottky diode shows that the current flow is easy when forward biased due to a lower barrier for electrons in the semiconductor. In reverse bias a saturation current is reached which is determined by the barrier for electrons in the metal, Φ_B. In silicon devices with large values of Φ_B, the reverse saturation current may be determined by the generation current in the depletion region. For those with a smaller value of Φ_B, the barrier height determines the reverse saturation current. In practical devices the barrier height is greatly influenced by the interface states.

Ohmic contacts are formed with metal on degenerately doped semiconductors. If the potential barrier is very thin, then carriers can tunnel from the metal to the semiconductor and vice versa. A metal-N^+-N^- contact also makes a good, ohmic contact.

The metal–semiconductor contact is primarily a majority carrier device with little if any minority carrier conduction. It is therefore a very fast switching device and has smaller capacitance than the p-n junction in forward bias. Similar discussions for metal on p-type semiconductors are applicable with the majority carriers being holes.

PROBLEMS

7.1 An ideal metal–semiconductor contact is made of a metal that has $\Phi_M = 4.75$ eV, and a semiconductor that has $\chi = 4.00$ eV with $n_i = 10^{10}/cm^3$, $N_D = 10^{16}/cm^3$, and $E_G = 1.00$ eV; $kT = 0.026$ eV.

(a) Calculate the barrier for electrons in the metal.

(b) Calculate V_{bi}, the barrier for electrons in the semiconductor.

(c) Calculate the value of the depletion region width at thermal equilibrium.

(d) Calculate the maximum electric field.

(e) Sketch the energy band diagram in thermal equilibrium.

7.2 The ideal M–S contact has a semiconductor with $\chi = 4$ eV, $n_i = 10^{10} cm^3$, $N_D = 10^{16}/cm^3$, $E_G = 1.2$ eV, and $kT = 0.026$ eV. Various metals are applied to the semiconductor with $\Phi_M = 4, 4.25, 4.5, 4.75$, and 5.0 eV.

(a) Calculate and sketch Φ_B vs. Φ_M

(b) Calculate and sketch qV_{bi} vs. Φ_M

(c) If $K_S = 12$ obtain an equation, with numerical coefficients, for the depletion region width vs. Φ_M if $V_A = 0$

(d) What is the probability that an electron will have an energy equal to Φ_B if $\Phi_M = 4.75$ eV?

7.3 The purpose of this problem is to show that the number of electrons that can get over a barrier is exponentially related to the barrier height. Consider the electrons in the metal as they see the barrier Φ_B. If the density of states function for the metal is $N_M E^{1/2}$ (#/cm^3-eV) and the Fermi function describes the probability of occupancy,

(a) First sketch the density of states function, next to it the Fermi function, and next to that their product. (Hint: see Chapter 2, Volume I).

(b) Indicate by shading the diagram of part (a) which electrons clear the barrier and derive a formula, left as an integral, for the number of electrons that can go over the barrier, n_ϕ. Assume $\Phi_B > 3\ kT$.

(c) Since the integral of part (b) is not easy to evaluate, let's assume that $N_M E^{1/2}$ is not a very fast-changing function and let it be a constant of

$$N_M (\Phi_B + E_{FM})^{1/2}$$

and derive the formula for n_ϕ.

(d) (optional) To check that (c) is a good approximation use

$$\int x^m e^{ax}\, dx = \frac{x^m e^{ax}}{a} - \frac{m}{a} \int x^{m-1} e^{ax}\, dx$$

to show that the first term is identical to part (c) and that there is an error term. Is the error large or small?

7.4 In Schottky clamped TTL logic (STTL) circuits the time to switch the transistors is reduced considerably by placing a Schottky barrier diode in parallel with the p-n junction of the collector to base. Remember that the Schottky diode has very little storage time effects compared to the p-n junction. The Schottky device shorts the forward current of the collector to base when the transistor is saturated.

The values of $I_0 = 10^{-15}$ and $n_1 = 1.01$ for the p-n junction while $I_S = 10^{-12}$ and $n_2 = 1.10$ for the Schottky device.

(a) If $V_A = 0.4$ volts calculate the ratio of the Schottky current to the p-n current. Note that the reverse recovery time of the p-n junction increases with current.

(b) At what value of V_A will both devices have equal current?

(c) Discuss the practical reasoning behind this problem.

7.5 A capacitance bridge was used to take the 1 KHz capacitance data on a Schottky diode as listed in the table. Area $= 100~\mu m \times 100~\mu m$ square metal pad.

(a) Determine V_{bi}.

(b) Determine N_D if assumed to be a uniformly doped semiconductor.

Table P7.5

V_A	C(pF)
−1	1.294
−2	1.002
−3	0.8469
−4	0.7469
−5	0.6756
−6	0.6215

7.6 The doping density of the semiconductor can also be determined from $1/C^2$ plots if the semiconductor is not uniformly doped. Assume $N_D(x)$ and derive the equation below from the slope of the $(1/C^2)$ plot. Hint: think of using incremental charge and voltage, dV_a.

$$N_D(x_d) = \frac{-2}{qK_S\varepsilon_0 \left[\dfrac{d(1/C^2)}{dV_A}\right]_{x=x_d}}$$

7.7 For the ideal barrier in silicon, let $N_D = 10^{19}/cm^3$ and $\Phi_M = 4.75$ eV.

(a) Is it a tunnel ohmic contact? Explain.

(b) Draw an energy band diagram.

(c) Draw the energy band diagram with the metal having a negative potential.

7.8 An ideal Schottky barrier diode on N-type silicon has light shining on the device such that carrier generation occurs in the silicon region.

(a) Draw an energy band diagram showing how the carriers flow if the metal is on the left. Is the current positive or negative through the device if the terminals are short-circuited?

(b) If the depletion region generation is ignored and only the bulk silicon considered, how will the light affect the open circuit voltage? Draw an energy band diagram.

7.9 A p-type silicon – metal contact is made with aluminum ($\Phi_M = 4.75$ eV). $N_A = 5 \times 10^{15}/cm^3$, and $kT = 0.026$ eV.

(a) Sketch to scale a thermal equilibrium band diagram for the ideal case.

(b) Comment on the barriers for

(i) electrons in the metal

(ii) electrons in the silicon

(iii) holes in the silicon

(iv) holes in the metal

(c) If positive voltage is applied to the semiconductor, show the carrier flux and energy band diagram.

Suggested Readings

W. H. Hayt and G. W. Neudeck, *Electronic Circuit Analysis and Design*. 2nd ed. Boston: Houghton Mifflin, 1983. Chapters 1 and 2 are a good review of circuit models and diode circuits.

J. L. Moll, *Physics of Semiconductors*. New York: McGraw-Hill, 1964. Chapter 7 contains much more detailed discussions of generation and recombination in the depletion region, quasi-Fermi levels, and graded junctions.

R. S. Muller and T. I. Kamins, *Device Electronics for Integrated Circuits*. 2nd ed. New York: John Wiley & Sons, 1986. An excellent discussion of metal–semiconductor contacts in Chapter 3, while Chapter 5 is good for *p-n* junctions.

B. G. Streetman, *Solid State Electronic Devices*. 2nd ed. Englewood Cliffs, N.J.: Prentice-Hall, 1980. Chapter 5 is a complete discussion of the diode from fabrication to charge storage. Chapter 6 presents the many applications of the *p-n* junction as light emitting, tunnel, solar, photo, and varactor diodes.

S. M. Sze, *Semiconductor Devices: Physics and Technology*. New York: John Wiley & Sons, 1985. Chapter 2 has an advanced discussion of generation–recombination and Chapter 5 discusses metal–semiconductor contacts.

Appendix A
Exercise Problems

EXERCISE 1.1

If the surface concentration is fixed at the solid solubility limit of boron in silicon of $N_0 = 1.8 \times 10^{20}$ atoms/cm³ and boron is diffused at 950 °C for 30 minutes ($D_{boron} = 3.0 \times 10^{-15}$ cm²/s at that temperature) into an n-type substrate, calculate the boron impurities at 0.05 μm and 0.1 μm.

Solution:

Since 30 minutes = $30 \times 60 = 1800$ s, and with a constant surface concentration, the distribution is a complementary error function.

$$N(0.05\,\mu, 1800) = 1.8 \times 10^{20} \text{ erfc}\left[\frac{0.05 \times 10^{-4}}{2\sqrt{3 \times 10^{-15} \times 1800}}\right]$$

$$N(0.05\,\mu, 1800) = 1.8 \times 10^{20} \text{ erfc}[1.0758] \quad \text{where erfc}[1.0758] = 0.12816$$

from Table 1.1 or Figure P1.1, therefore

$$N(0.05\,\mu, 1800) = 1.8 \times 10^{20} \text{ erfc}[1.0758] = 2.3068 \times 10^{19}/\text{cm}^3$$

Similarly,

$$N(0.1\,\mu, 1800) = 1.8 \times 10^{20} \text{ erfc}[2.152] = 4.2169 \times 10^{17}/\text{cm}^3$$

Note that the impurities have decreased by two orders of magnitude while the distance has only doubled.

EXERCISE 2.1

For the energy band diagram shown:
(a) Sketch the charge density, electric field, and potential.
(b) Let $kT = 0.026$ eV and calculate the maximum electric field and the built-in potential, V_{bi}.

Solution:

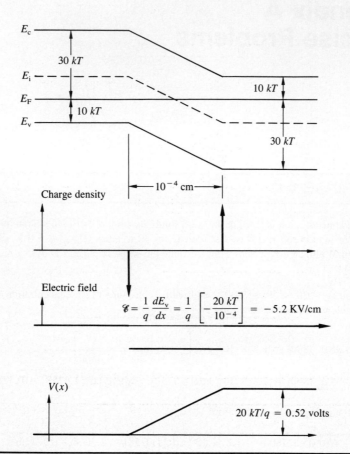

EXERCISE 2.2

Sketch the energy band diagram (in units of kT), charge density, and potential for a silicon n^+-p junction with $N_D = 2 \times 10^{+17}/\text{cm}^3$ and $N_A = 5 \times 10^{+15}/\text{cm}^3$. Let $kT = 0.026$ eV or $kT/q = 0.026$ volts and $n_i = 10^{+10}/\text{cm}^3$.

Solution:

$$V_{bi} = 0.026 \ln\left[\frac{10^{33}}{10^{20}}\right] = 29.934\,kT = 0.77827 \text{ volts}$$

$$E_F - E_i = kT \ln\left[\frac{2 \times 10^{17}}{10^{10}}\right] = 16.811\,kT$$

$$E_F - E_i = -kT \ln\left[\frac{5 \times 10^{15}}{10^{10}}\right] = -13.122\,kT \quad \text{and} \quad E_G = 43.08\,kT$$

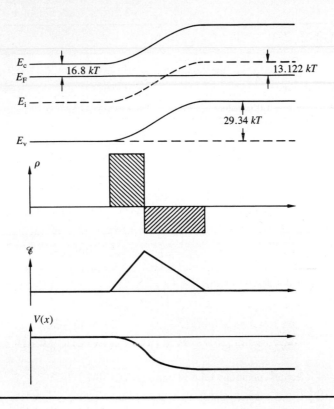

EXERCISE 2.3

For the example at the end of Section 2.3, calculate the junction voltage required to make W equal to

(a) 2 μm

Solution:

$$2 \times 10^{-4} = \left[1.306 \times 10^7 (0.659 - V_A) \frac{1.1 \times 10^{16}}{10^{31}} \right]^{1/2}$$

$$0.659 - V_A = 2.7844 \quad \text{and therefore } V_A = -2.1254 \text{ volts}$$

(b) 0.6 μm

Solution:

$$0.659 - V_A = \frac{0.36 \times 10^{-8}}{1.306 \times 10^7 (1.1 \times 10^{-15})} = 0.25059 \quad \text{and} \quad V_A = 0.40841$$

EXERCISE 3.1

A silicon step junction has $N_A = 10^{+16}/cm^3$ and $N_D = 10^{+15}/cm^3$ with $n_i = 10^{+10}/cm^3$ and $kT = 0.026$ eV. Calculate

(a) $\Delta p_n(x_n)$ if $V_A = 0.4$ and 0.6 volts

Solution:

$$n_{p0} = \frac{n_i^2}{N_A} = \frac{10^{20}}{10^{16}} = 10^4/cm^3$$

$$p_{n0} = \frac{n_i^2}{N_D} = \frac{10^{20}}{10^{15}} = 10^5/cm^3$$

$$\Delta p_n(x_n) = 10^5(e^{0.4/0.026} - 1) = 4.581 \times 10^{11}/cm^3$$

$$\Delta p_n(x_n) = 10^5(e^{0.6/0.026} - 1) = 1.052 \times 10^{15}/cm^3$$

(b) $\Delta n_p(-x_p)$ if $V_A = 0.4$ and 0.6 volts

Solution:

$$\Delta n_p(-x_p) = 10^4(e^{0.4/0.026} - 1) = 4.581 \times 10^{10}/cm^3$$

$$\Delta n_p(-x_p) = 10^4(e^{0.6/0.026} - 1) = 1.052 \times 10^{14}/cm^3$$

(c) Are low-level injection conditions valid in parts (a) and (b)?

In part (a) at $V_A = 0.6$ it is not valid since $\Delta p_n(x_n)$ is not $\ll 10^{15}/cm^3$. All the other cases are valid solutions.

EXERCISE 3.2

A step junction has $N_A = 10^{+17}/cm^3$ and $N_D = 5 \times 10^{+15}/cm^3$ with $D_N = 30$ cm^2/s, and $D_P = 12$ cm^2/s. Let $n_i = 10^{+10}/cm^3$, $kT = 0.026$ eV, $L_N = 10 \times 10^{-4}$ cm, $L_P = 15 \times 10^{-4}$ cm, and $A = 10^{-4}/cm^2$. Calculate:

(a) the ratio of the hole current to the total current in the depletion region.

Solution:

$$I_p(x_n) = qAn_i^2\left[\frac{D_P}{L_P}\frac{1}{N_D}\right](e^{qV_A/kT} - 1)$$

$$I(x_n) = qAn_i^2\left[\frac{D_P}{L_P}\frac{1}{N_D} + \frac{D_N}{L_N}\frac{1}{N_A}\right](e^{qV_A/kT} - 1)$$

$$\frac{I_p}{I} = \frac{1}{1 + \dfrac{D_N}{D_P}\dfrac{L_P}{L_N}\dfrac{N_D}{N_A}} = \frac{1}{1 + \dfrac{30}{12}\dfrac{15}{10}\dfrac{5 \times 10^{15}}{10^{17}}} = \frac{1}{1 + 0.1875} = 0.8421$$

or 84.2% of the total.

(b) Repeat part (a) for the electron current.

Solution:

$$\frac{I_n}{I} = \frac{1}{1 + \dfrac{D_P}{D_N}\dfrac{L_N}{L_P}\dfrac{N_A}{N_D}} = \frac{1}{1 + \dfrac{12}{30}\dfrac{10}{15}\dfrac{10^{17}}{5 \times 10^{15}}} = 0.15789$$

or 15.8% of the total.

(c) If N_D were made smaller, discuss how this would affect the ratios of parts (a) and (b).

As N_D decreases the hole current percentage will increase and the electron current will decrease their percentage of the total.

EXERCISE 3.3

For the data given in Exercise 3.2 calculate the following at $V_A = 0.4$:

(a) $\Delta p_n(x' = 0)$ and $\Delta p_n(x' = 60 \ \mu\text{m})$

$$\Delta p_n(x') = p_{n0}(e^{qV_A/kT} - 1)e^{-x'/L_P} = \frac{n_i^2}{N_D}(e^{10.4/0.026} - 1)e^{-x'/L_P} = 9.6047 \times 10^{10}e^{-[x'/(15\times 10^{-4})]}$$

$$\Delta p_n(0') = 9.6047 \times 10^{10}/\text{cm}^3 \quad \text{and} \quad \Delta p_n(x' = 60 \times 10^{-4}) = 1.759 \times 10^9/\text{cm}^3$$

(b) $I_P(x' = 0)$ and $I_P(x' = L_P)$

$$I_P(x') = qAn_i^2\left[\frac{D_P}{L_P}\frac{1}{N_D}\right](e^{qV_A/Kt} - 1)e^{-x'/L_P} = 1.024 \times 10^{-16}(e^{15.385} - 1)e^{-x'/(15\times 10^{-4})}$$

$$I_P(x' = 0) = 4.9195 \times 10^{-10} \quad \text{and} \quad I_P(x' = L_P) = 4.9195 \times 10^{-10}e^{-1} = 1.8098 \times 10^{-10}$$

(c) If the recombination rate were, for some reason, to quadruple in the n-region, calculate the new value of $I_P(x' = 0)$.

τ_p becomes $\tau_p/4$ and hence L_P becomes $L_P/2$, and since $I_P(x')$ is proportional to $1/L_P$ it will double. Therefore $I_P(x' = 0)$ will double to 9.839×10^{-10}.

EXERCISE 4.1

For a silicon p-n step junction doped at $N_A = 10^{+17}/\text{cm}^3$ and $N_D = 5 \times 10^{+15}/\text{cm}^3$ with $n_i = 10^{+10}/\text{cm}^3$, $kT/q = 0.026$ volts, $m = 4.0$, and $\mathscr{E}_{CR} \cong 4 \times 10^5$ V/cm:

(a) Calculate V_{BR}.

$$V_{BR} \cong (4 \times 10^5)^2 \left(\frac{8.854 \times 10^{-14} \times 11.8}{2q} \right) \frac{(10^{17} + 5 \times 10^{15})}{10^{17} \times 5 \times 10^{15}} = 109.7 \text{ volts}$$

(b) At what value of V_A will the current have increased by
  (i) 10,
  (ii) 100?

$$M = 10 = \frac{1}{1 - \left(\frac{|V_A|}{V_{BR}}\right)^4} \quad \text{or} \quad \left(\frac{|V_A|}{V_{BR}}\right)^4 = \frac{M-1}{M}$$

$$V_A = V_{BR} \sqrt[4]{\frac{M-1}{M}} = 109.7 \sqrt[4]{\frac{10-1}{10}} = 106.85$$

$$V_A = 109.7 \sqrt[4]{\frac{100-1}{100}} = 109.42$$

(c) What is the ratio of $I_{G-R}(V_A = -20)/I_{G-R}(V_A = -2)$?

$$V_{bi} = 0.7603 \quad \text{and} \quad W = K(V_{bi} - V_A)^{1/2}$$

$$\frac{I_{R-G}(-20)}{I_{R-G}(-2)} = \left[\frac{0.7603 + 20}{0.7603 + 2}\right]^{1/2} = 2.7424$$

EXERCISE 4.2

For a p^+-n step junction that has $W = 10^{-4}$ cm, $n_i = 10^{+10}/\text{cm}^3$, $N_D = 5 \times 10^{+15}/\text{cm}^3$, $A = 10^{-4}/\text{cm}^2$, $kT = 0.026$ eV, $\tau_0 = 20$ μs, and $D_P/L_P = 10^4$ with $n_1 = 1.05$ and $n_2 = 2.0$, what is:

(a) The ratio of $I_{Rec}/I_{diffusion}$

$$I_{diffusion} \cong qAn_i^2 \left[\frac{D_P}{L_P} \frac{1}{N_D}\right] (e^{qV_A/1.05kT} - 1)$$

$$I_{Rec} = \frac{qAn_i W}{2\tau_0} (e^{qV_A/2.00kT} - 1)$$

$$\frac{I_{Rec}}{I_{diffusion}} = \frac{4 \times 10^{-13}(e^{qV_A/2.00kT} - 1)}{3.2 \times 10^{-15}(e^{qV_A/1.05kT} - 1)} = \frac{125(e^{19.231V_A} - 1)}{(e^{36.63V_A} - 1)}$$

(b) The value of the ratio at $V_A = 0.05$, $V_A = 0.1$, and $V_A = 0.2$? From above,

$$\frac{I_{Rec}}{I_{diffusion}} = \frac{125(e^{19.231 \times 0.05} - 1)}{(e^{36.63 \times 0.05} - 1)} = 38.519$$

$$\frac{I_{\text{Rec}}}{I_{\text{diffusion}}} = \frac{125(e^{19.231\times0.1} - 1)}{(e^{36.63\times0.10} - 1)} = 19.228$$

$$\frac{I_{\text{Rec}}}{I_{\text{diffusion}}} = \frac{125(e^{19.231\times0.2} - 1)}{(e^{36.63\times0.2} - 1)} = 3.7717$$

EXERCISE 5.1

A p^+-n step junction is doped $N_A = 10^{+16}/\text{cm}^3$ and $N_D = 10^{+15}/\text{cm}^3$. Let $kT = 0.026$ eV, $n_i = 10^{+10}/\text{cm}^3$, $A = 10^{-4}$ cm^2, $L_N = 14 \times 10^{-4}$ cm, $L_P = 35 \times 10^{-4}$ cm, $D_N = 20$ cm^2/s, and $D_P = 12.5$ cm^2/s.

(a) Calculate the depletion capacitance if $V_A = 0$ and -2 volts.

$$V_{\text{bi}} = 0.6585 \quad \text{and} \quad W = \sqrt{1.306 \times 10^7(0.6585)\left[\frac{1.1 \times 10^{16}}{10^{31}}\right]} = 0.9727 \times 10^{-4} \text{ cm}$$

$$C_{j0} = \frac{K_S\varepsilon_0 A}{W} = \frac{11.8 \times 8.854 \times 10^{-14}10^{-4}}{0.9727 \times 10^{-4}} = 1.074 \text{ pF}$$

$$C_j(V_A = -2) = \frac{1.074 \text{ pF}}{\left[1 - \frac{(-2)}{0.6585}\right]^{1/2}} = 0.5345 \text{ pF}$$

(b) Calculate the low-frequency conductance at $V_A = 0$ and -2 volts.

$$G_0 = \frac{q(I + I_0)}{kT} = \frac{q}{kT}\left\{qAn_i^2\left[\frac{D_P}{L_P}\frac{1}{N_D} + \frac{D_N}{L_N}\frac{1}{N_A}\right](e^{qV_A/kT})\right\}$$

$$G_0 = \frac{8 \times 10^{-15}}{0.026}e^{qV_A/kT}$$

$$G_0 = 3.077 \times 10^{-13} \quad \text{at } V_A = 0$$

$$G_0 = 3.077 \times 10^{-13}e^{-2/0.026} = 1.2046 \times 10^{-46}$$

EXERCISE 5.2

If for a p-n junction $j\omega\tau_p = 1$ and $j\omega\tau_n = 0.90$ at $\omega = 10^8$ rad/s with $A = 10^{-4}$ cm^2, $V_A = 0.6$ volts, and

$$\frac{qD_N}{L_N}p_{n0} = 5.0 \times 10^{-11} \quad \text{and} \quad \frac{qD_P}{L_P}n_{p0} = 1.6 \times 10^{-10}$$

Calculate G and C_D.

$$Y = \frac{qA}{kT}\left[1.6 \times 10^{-10}\sqrt{1 + j1} + 5.0 \times 10^{-11}\sqrt{1 + j0.90}\right]e^{0.6/0.026}$$

$$Y = 8.647 \times 10^5[1.6 \times 10^{-10}(1.414\angle 45°)^{1/2} + 5 \times 10^{-11}(1.345\angle 41.99°)^{1/2}]$$

$$Y = 8.647 \times 10^5[2.995 \times 10^{-10} + j0.9359 \times 10^{-10}]$$

$$Y = 1.9884 \times 10^{-4} + j0.80927 \times 10^{-4}$$

$$G = 1.9884 \times 10^{-4} \quad \text{and} \quad C_D = \frac{0.80927 \times 10^{-4}}{10^8} = 0.80927 \text{ pF}$$

EXERCISE 6.1

The p^+-n junction has $I_F = 1$ ma and $\tau_p = 1$ μs and is turned off by $I_R = 0$. Derive an equation for $Q_P(t)$.

$$\frac{dQ_P(t)}{dt} = 0 - \frac{Q_P(t)}{\tau_p}$$

$$\frac{dQ_P(t)}{Q_P(t)} = -\frac{dt}{\tau_p}$$

$$\ln[Q_P]_{Q_P(0)}^{Q_P(t)} = -\left[\frac{t}{\tau_p}\right]_0^t$$

$$Q_P(t) = Q_P(0)e^{-t/\tau_p}$$

EXERCISE 6.2

For the turn-on transient by a constant current, derive a formula for the time required to reach 90% of the total current.

At infinite time, $v_A = V_A$ and 90% is $0.9\, kT/q$ $(\ln[I_F/I_0])$. In Eq. (6.21b) the "-1" term is very small compared to the exponent term as t approaches infinity. Then Eq. (6.22) becomes

$$0.9V_A \cong \frac{kT}{q} \ln\left[\frac{I_F}{I_0}(1 - e^{-t/\tau_p})\right] = 0.9\frac{kT}{q} \ln\left[\frac{I_F}{I_0}\right]$$

$$\frac{I_F}{I_0}(1 - e^{-t/\tau_p}) = \left(\frac{I_F}{I_0}\right)^{0.9}$$

$$t = t_p \ln\left[\frac{1}{1 - \left(\frac{I_F}{I_0}\right)^{0.1}}\right]$$

EXERCISE 7.1

A metal with $\Phi_M = 4.75$ eV (Au) and a semiconductor (Si) with $\chi = 4.05$ eV are formed into an ideal metal–semiconductor contact. If $kT = 0.026$ eV, $n_i = 10^{10}/cm^3$ and $N_D = 10^{+16}/cm^3$

(a) Is it a Schottky barrier or an "ohmic" contact?

If $\Phi_M > \chi + (E_c - E_{FS})_{bulk}$ then it is a Schottky diode

$$(E_{FS} - E_i) = kT \ln\left[\frac{N_D}{n_i}\right] = (0.26) \ln\left[\frac{10^{16}}{10^{10}}\right] = 0.3592 \text{ eV}$$

$$(E_c - E_{FS})_{bulk} = 0.56 - 0.3592 = 0.2008 \text{ eV}$$

Hence, $\Phi_S = 4.05 + 0.2008 = 4.2508$ eV which is less than 4.75 eV $= \Phi_M$.

(b) Calculate the ideal barrier height for an electron in the metal at E_{FM}.

$$\Phi_M - \chi = 4.75 - 4.05 = 0.700 \text{ eV} = \Phi_B$$

(c) Calculate the ideal barrier height for an electron at E_c in the semiconductor.

$$qV_{bi} = \Phi_M - \chi - (E_c - E_{FS})_{bulk}$$

$$= 4.75 - 4.05 - 0.2008 = 0.4992 \text{ eV}$$

$$V_{bi} = 0.4992 \text{ volts}$$

(d) Calculate x_n for $V_A = 0$ and $V_A = -2$ volts

$$x_n = \sqrt{\frac{2K_S\varepsilon_0(V_{bi} - V_A)}{qN_D}}$$

$$x_n = \sqrt{\frac{2(11.8)8.85 \times 10^{-14}}{1.6 \times 10^{-19} \times 10^{16}}(0.4992 - V_A)}$$

$$x_n = 36.116 \times 10^{-6}\sqrt{0.4992 - V_A} \qquad @ V_A = 0 \qquad x_n = 25.57 \times 10^{-6}$$

$$x_n = 0.25517 \times 10^{-4}$$

$$x_n = 0.25517 \ \mu m$$

$@ V_A = -2$

$$x_n = 57.095 \times 10^{-6} = 0.57095 \times 10^{-4} = 0.57095 \ \mu m$$

EXERCISE 7.2

A Schottky barrier diode has a value of $\Phi_B = 0.5$ eV, $I_S = 5 \times 10^{-12}$ A, $n = 1.07$, and $kT = 0.026$ eV. If a second device were to be made with $\Phi_B = 0.7$ eV with everything else the same, what is the current through both if $V_A = 0.4$ volts and $V_A = -2$ volts?

$$I_1 = 5 \times 10^{-12}(e^{0.4/1.07(0.026)} - 1) = 5 \times 10^{-12}(e^{35.945 - V_A} - 1)$$

$$I_1 = 8.7765 \ \mu A \qquad @ \ V_A = 0.4 \qquad I_1 \cong x \times 10^{-12} \ A \qquad @ \ V_A = -2$$

since

$$I_{S1} = KAe^{-\Phi_B/kT} = KAe^{-0.5/0.026} = 5 \times 10^{-12} \qquad \text{then } KA = 1.1241 \times 10^{-3}$$

$$I_{S2} = 1.1241 \times 10^{-3}e^{-0.7/0.026} = 2.2816 \times 10^{-15} \ A$$

$$I_2 = 2.2816 \times 10^{-15}(e^{35.945V_A} - 1)$$

$$= 4.0049 \times 10^{-9} \ A \qquad @ \ V_A = 0.4 \ \text{volts}$$

$$I_2 \cong 2.2816 \times 10^{-15} \qquad @ \ V_A = -2 \ \text{volts}$$

Appendix B
Volume Review Problem Sets and Answers

To work the following problems often requires an integrated knowledge of the subjects presented in Chapters 1 through 7. These problems serve as a general review of the subject matter.

PROBLEM SET A

A.1 List two methods used for fabricating an abrupt junction and two methods for a graded junction.

A.2 The thermal diffusion of impurities has a fixed number of the impurities. How do you expect the impurities to distribute themselves into silicon? Where is the metallurgical junction?

A.3 Why are regions I, II, and III nonideal?

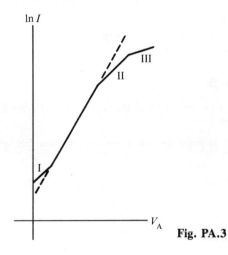

Fig. PA.3

A.4 Explain the phenomena of "Zener" and "avalanche."

A.5 Sketch an energy band diagram for a p-n junction and explain what the "built-in potential" is and how it changes with increased doping of the acceptor material.

A.6 Explain *how* generation in the depletion region affects the $I - V_A$ characteristic in reverse bias.

A.7 For an abrupt p-n junction, sketch the minority carrier concentrations for forward bias, where $N_D > N_A$. Label *all* the boundary conditions and excess carrier concentrations, etc.

A.8 If N_D increases ($\uparrow$), how will the following parameters vary for a p^+-n junction?

$$\tau_p \qquad \tau_n \qquad C_J \qquad C_D \qquad \mathscr{E}(0)$$

A.9 A p-n junction is such that $L_P \gg W_N$ and $L_N \gg W_P$, and

$$\Delta p_n(W_N) = 0$$
$$\Delta n_p(-W_P) = 0$$

Sketch the minority carrier distributions for forward and reverse bias.

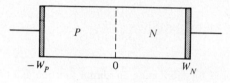

Fig. PA.9

A.10 A p-n^+ junction is forward biased. Sketch the junction's conductance and diffusion capacitance vs. $\omega\tau_n$.

A.11 A p-n^+ junction is forward biased at I_F when open circuited at $t = 0$. Sketch the minority carrier distribution $\Delta n_p(x, t)$. Also sketch $v_A(t)$.

A.12 Explain why switching of the diode from a large reverse bias to $V_A = 0$ is quite rapid.

PROBLEM SET B

B.1 The n-p junction diode is doped so that $N_D = 5N_A$. The diode is at thermal equilibrium. *Label* all significant points.

(a) Sketch (roughly to scale) the charge-density diagram. Let $|qN_A|$ equal one unit.

If $E_G = 4$ units on the graph:

(b) sketch the electron energy band diagram roughly to scale. Let $|E_F - E_v|$ on the p-side be one unit and $|E_c - E_F|$ on the n-side be one-half unit.

(c) Sketch the electric field.

(d) Sketch the potential.

(e) What is the value of the built-in potential (or contact potential), in # of squares?

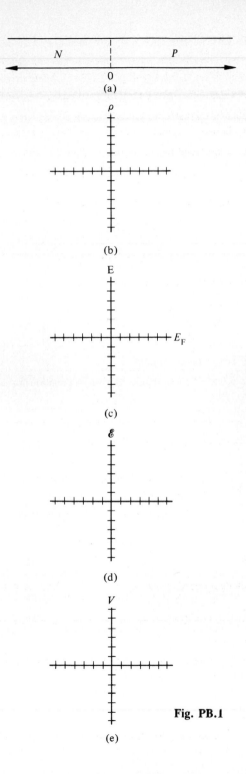

Fig. PB.1

B.2 List several reasons why the calculated I_0 of a reverse-biased diode does not agree with the measured value.

B.3 A p-n abrupt junction is symmetrically doped ($N_A = N_D$) and driven from a current source that is stepped from zero to I_F at $t = 0$.

(a) Sketch $Q_N(t)$ and $Q_P(t)$.

(b) Sketch $n_p(x, t)$ and $p_n(x, t)$.

(c) Explain how charge is increased and/or decreased from $t = 0$ to $t = \infty$.

B.4 If $Y = 10^{-3}\sqrt{1 + j\omega 10^{-7}}$ for a p^+-n junction, calculate G and C_D for $\omega = 10^7$ rad/sec and 5×10^7 rad/sec.

B.5 Forward bias diode:

$n_i^2 = 10^{20}/\text{cm}^3$,	$\tau_n = 10^{-7}$ sec,
$N_A = 10^{16}/\text{cm}^3$,	$\tau_p = 5 \times 10^{-7}$ sec,
$N_D = 10^{14}/\text{cm}^3$,	$D_P = 20 \text{ cm}^2/\text{sec}$,
$D_N = 50 \text{ cm}^2/\text{sec}$, $\Delta n_p(-W_N) = 0$,	
$W_N = 2 \times 10^{-4}$ cm	

(a) Sketch the minority carrier concentrations.

(b) If no recombination occurs in the p-region, derive an equation for $\Delta n_p(x)$.

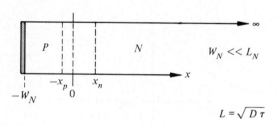

Fig. PB.5

B.6 Will the work function of a semiconductor change with doping? Explain.

B.7 For a Schottky diode will V_{bi} increase or decrease with increased doping?

B.8 If the doping in a Schottky barrier diode is quadrupled, will the depletion width be halved? Explain.

B.9 A Schottky barrier diode has $I_0 = 10^{-14}$ amps at $V_A = 0$. Due to barrier lowering of 0.05 eV, calculate the new value of I_0.

B.10 What are the two types of "ohmic" contacts to n-type silicon and under what conditions do they occur?

ANSWERS TO SET A REVIEW PROBLEMS

A.1   Alloy   Diffusion

       $\updownarrow$ $\updownarrow$

  Epitaxial Ion implant

A.2 Gaussian, $N(x) = N_B$.

A.3 I, recombination in W; II, high injection; III, R_s.

A.4 Zener is tunneling; avalanche is where the carrier ionizes a silicon atom by impact collision to free an electron–hole pair.

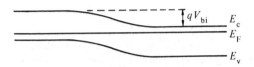

Fig. AA.5

A.5 V_{bi} increases with N_A or N_D.

A.6 Adds carriers generated in W to the current flow; reverse current is increased and gets larger with V_A.

A.7

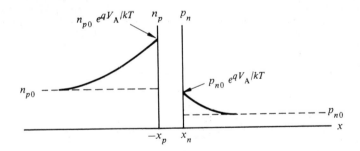

Fig. AA.7

A.8 τ_p not affected, τ_n not affected, C_J and $\mathscr{E}(0)$ increase, C_D decreases.

A.9

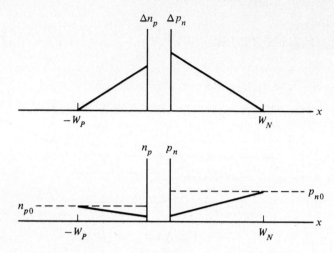

Fig. AA.9

A.10

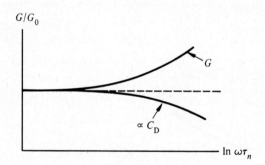

Fig. AA.10

A.11

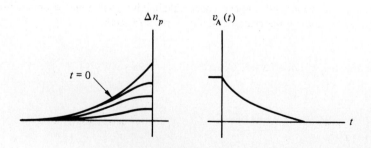

Fig. AA.11

A.12 Because the depletion width change is the movement of majority carriers that change at about the dielectric relaxation time.

ANSWERS TO SET B REVIEW PROBLEMS

B.1

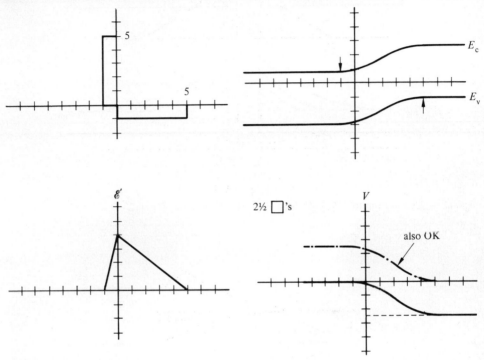

Fig. AB.1

B.2 Generation in W, surface leakage

B.3 Slope at $x = 0$ is constant! I_F injects charge and builds up with time; recombination limits the buildup and at $t =$ infinity the charge entering $=$ recombination.

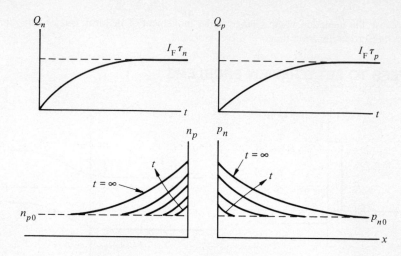

Fig. AB.3

B.4 $G = 1.0986$ m mho, $C_D = 45.5$ pF; $G = 1.746$ m mho. $C_D = 28.6$ pF.

B.5

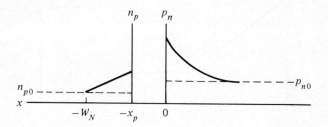

Fig. AB.5

$$t_n \to \infty \therefore D_N \frac{d^2 \Delta n_p}{dx^2} = 0$$

$$\Delta n_p(x) = n_{p0}(e^{qV_A/kT} - 1)\frac{x + W_N}{W_N - x_p}$$

B.6 Yes, since $\Phi_S = \chi + (E_c - E_F)_{bulk}$, as the doping increases $E_c - E_F$ decreases and hence Φ_S decreases.

B.7 V_{bi} increases since Φ_S decreases with larger doping; $V_{bi} = \Phi_M - \Phi_S$.

B.8 No, x_n decreases due to $1/\sqrt{N_D}$, but remember V_{bi} increases, so therefore by less than $1/2$.

B.9

$$AKe^{-\Phi_B/kT} = 10^{-14} \quad \text{and} \quad AKe^{-(\Phi_B-0.05)/kT} = I_{02}$$

by taking the ratio results in

$$I_{02} = 10^{-14}e^{0.05/0.026} = 6.842 \times 10^{-14}$$

B.10 $\Phi_M < \Phi_S$ leads to a "negative V_{bi}" contact (no barrier). $\Phi_M > \Phi_S$ with N^+ silicon leads to a tunnel contact.

Appendix C
List of Symbols

a	grading constant, linearly graded ($\#/\text{cm}^4$)
A	cross-sectional area (cm^2)
C	capacitance (F)
C_J	depletion capacitance
C_{J0}	depletion capacitance at $V_A = 0$
C_0	depletion capacitance of M$-$S diode (F) at $V_A = 0$
C_D	diffusion capacitance
D	impurity diffusion constant (cm^2/sec)
D_N	electron diffusion constant (cm^2/sec)
D_P	hole diffusion constant (cm^2/sec)
E_c	conduction band edge energy level (eV)
E_F	Fermi energy level (eV)
E_{FM}	Fermi energy in metal
E_{FS}	Fermi energy in semiconductor
E_i	intrinsic Fermi energy level (eV)
E_0	vacuum energy level (eV)
E_v	valence band edge energy level (eV)
E_{cn}	conduction band edge in n-material
E_{cp}	conduction band edge in p-material
E_{vn}	valence band edge in n-material
E_{vp}	valence band edge in p-material
G	generation rate for electrons and holes ($\#/\text{cm}^3$-sec)
G	conductance $(\Omega)^{-1}$
G_0	low-frequency conductance
i	diode signal current

I	total current (A)
I_N	current due to electrons (A)
I_P	current due to holes (A)
I_0	diode reverse saturation current (A)
I_{R-G}	recombination-generation current component (A)
J	total current density (A/cm^2)
J_{M-S}	current density, metal to semiconductor (A/cm^2)
J_N	electron current density (A/cm^2)
J_{S-M}	current density, semiconductor to metal (A/cm^2)
J_0	reverse saturation current density for M–S
J_N	electron current density (A/cm^2)
J_P	hole current density (A/cm^2)
k	Boltzmann's constant
K_0	relative dielectric constant of oxide
K_S	relative dielectric constant of semiconductor
kT	thermal energy (eV)
L_N	electron minority carrier diffusion length (cm)
L_N^*	complex diffusion length, electrons
L_P	hole minority carrier diffusion length
L_P^*	complex diffusion length for holes
M	multiplication factor
m	exponent in multiplication
m	depletion capacitance exponent, $1/3 \leqq m \leqq 1/2$
N_A	acceptor impurity concentration (#/cm^3)
N_D	donor impurity concentration (#/cm^3)
$N(x,t)$	impurity concentration (#/cm^3)
N_0	surface concentration (#/cm^3)
N_A^-	ionized acceptor concentration (#/cm^3)
N_D^+	ionized donor concentration
n	ideality factor for a diode
n_1	ideality factor for diffusion currents
n_2	ideality factor for recombination currents
n_0	electron concentration at thermal equilibrium
n_n	electron concentration in n-material (#/cm^3)
n_p	electron concentration in p-material (#/cm^3)
n	electron concentration (#/cm^3)
n_{n0}	electron concentration in n-material at thermal equilibrium

n_{p0}	electron concentration in p-material at thermal equilibrium
n_i	intrinsic carrier concentration ($\#/\text{cm}^3$)
n_S	electron concentration at M$-$S interface
n^+	degenerately doped n-type material
p^+	degenerately doped p-type material
p_p	hole concentration in p-material ($\#/\text{cm}^3$)
p_n	hole concentration in n-material ($\#/\text{cm}^3$)
p	hole concentration ($\#/\text{cm}^3$)
p_{n0}	hole concentration in n-material at thermal equilibrium
p_{p0}	hole concentration in p-material at thermal equilibrium
p_0	hole concentration at thermal equilibrium ($\#/\text{cm}^3$)
$\tilde{p}_N(x, t)$	signal component of hole concentration
$\hat{p}_N(x)$	x variation of hole signal component
Q	total number of impurities initially deposited at the surface
Q_M	surface charge density in metal (coul/cm^2)
Q_p	hole minority carrier charge (coul)
Q_S	total charge in semiconductor (coul/cm^2)
q	charge on an electron (1.602×10^{-19} coul)
R_p	projected range (cm)
R_s	diode series resistance (bulk and contact) (Ω)
r	dynamic (small signal) resistance
t	time (sec)
T	temperature in degrees Kelvin ($^\circ$K)
t_s	diode storage time
t_{rr}	reverse recovery time
t_r	recovery time
V_A	diode applied voltages (V)
V_{bi}	junction built-in voltage (V)
$V(x)$	potential (V)
V_N	contact potential of n-material to metal (V)
V_P	contact potential of p-material to metal (V)
V_j	junction voltage across depletion region (V)
V_{BR}	diode breakdown (avalanche or Zener) voltage
v_A	total instantaneous applied voltage
V_A	dc value of applied voltage
v_a	small signal applied voltage
W	depletion region total width (cm)

x_n	n-region depletion region width (cm)
x_p	p-region depletion region width (cm)
x_j	junction depth (cm)
x, x', x''	x-axis variable (cm)
Y	admittance $(1/\Omega)$
Y_P	admittance due to minority carrier holes in n-region
Y_N	admittance due to minority carrier electrons in p-region
∇	gradient symbol
$\mathscr{E}$	electric field (V/cm)
$\mathscr{E}_{CR}$	electric field at avalanche (V/cm)
$\mathscr{E}_{max}$	maximum electric field (V/cm)
ε_0	permittivity of free space $[8.854 \times 10^{-14}$ (farad/cm)$]$
μ_n	electron mobility (cm^2/V-sec)
μ_p	hole mobility (cm^2/V-sec)
ρ	charge density (coul/cm^3)
τ_n	electron minority carrier lifetime (sec)
τ_p	hole minority carrier lifetime (sec)
τ_0	effective lifetime in depletion region
ϕ	implant dose (#/cm^2)
Φ	work function (eV)
Φ_B	M–S energy barrier (eV)
Φ_M	work function for a metal (eV)
Φ_S	work function for a semiconductor (eV)
χ	electron affinity (eV)
ω	radian frequency (rad/sec)
Δn_p	injected electron concentration in p-material
Δn_n	injected electron concentration in n-material
Δp_n	injected hole concentration in n-material
Δp_p	injected hole concentration in p-material
$\overline{\Delta p_n}(x)$	the dc value of excess hole concentration
ΔR_p	straggle (cm)

Index

a-c equivalent circuit. *See* Models
Abrupt junction, 3. *See also* Step junction
Admittance, junction, high frequency, 93–107
  junction, high frequency, forward bias,
    100–107
  low frequency, forward bias, 98–99
  p^+n, 105–107
  pn^+, 108
  small signal, 93
  reverse bias, 93–100
Alloyed junction, 3–4
Aluminum (Al), silicon, 3–4
Aluminum oxide, 5
Applied voltage, 35
Avalanche breakdown, 75–79

Boundary conditions:
  depletion edge, 58–60
  reverse recovery time, 114–115
Breakdown voltage, 75–80
Breakdown, Zener, 79–80
Built-in potential, 19, 24, 26–28,
  graded, 41–42
  metal-semiconductor, 129, 132
Bulk region, 19

Capacitance, depletion, 94–98
  diffusion, 104–107
  frequency dependence, 106–107
  metal-semiconductor, 140–142
  voltage variable, 95
Carrier concentration, equilibrium, 20–22

high-level injection, 85–86
  junction, forward bias, 62
  junction, reverse bias, 62
  step junction, 20–22
  storage time, 111
  thermal equilibrium, 48
Carrier flux, forward bias, 49–50
  reverse bias, 51–52
  thermal equilibrium, 46
Charge density, 22
  graded junction, 40
Charge neutrality, 20
Charge storage, step junction, 111–114
  turn-on, 118–119
Compensation, 6
Complementary error function, 6–7
  evaluation, 14–15
Conductance, 93
  forward bias, 99, 104
  high frequency, 104
  low frequency, 99, 105
  reverse bias, 99
Conduction band, 24
Contact potential, 35
Critical electric field, 76
  at breakdown, 77
  doping dependence, 78
Current, depletion region, 56–57
  diffusion, electron, 20–21
  drift, electron, 20–21
  forward, 84–85
  generation in W, 80–82
  recombination in W, 84–85